TIDES

Isidore Okpewho

Longman Group UK Limited,
Longman House, Burnt Mill, Harlow,
Essex CM20 2JE, England
and Associated Companies throughout the world.

First published 1993

Set in 10/11pt Baskerville

Produced by Longman Singapore Publishers Pte Ltd
Printed in Singapore

ISBN 0 582 10276 6

Acknowledgement

We are grateful to the author, J.P. Clark Bekederemo, for permission to reproduce his poem 'Tide-Wash' from A REED IN THE TIDE pubd. Longman Group UK Ltd 1965.

Up the laughing stream
We raced down the sun.
Who there thought such fun
Could end? We held one steam.

But the pulse that never
Gave sign fell through the sand;
Depleted now we stand
Exposed more than ever.

John Pepper Clark

This story is entirely fictitious, and is in no way intended as a mirror of any events past, present or future.

1
RIPPLE

Row, row the boat
My brother
Row, row the boat
It's howling in the east
It's howling in the west
Oh, brother
Let's row it on to Ezebiri
Before we're drenched in rain

Ijo folk song

For my sisters

Kanwulia *and* Juliet

Acknowledgement

I would like to thank Dr J. Egbe Ifie for making the pieces of Ijo folk song found within the following pages available to me. He is, of course, in no way accountable for the use to which they have been put.

Lagos, 25.8.76

Dear Tonwe,

I could have sworn I'd lost you after you left Lagos! I knew that even though you took the recent mass retirements which cost us our jobs with the philosophical calm for which you're well known, you were bound to be very angry like the rest of us. So I wasn't surprised to learn you'd gone home. But hardly did I suspect that home meant your little village on the Forcados river – I assumed you'd gone to Warri or perhaps Port Harcourt (I understand the new boundary adjustments put half of your village in the Rivers State!) and taken up some business or other. But to have actually retired to Seiama and settled down to farming and fishing – I swear, I never could have associated a refined man like you with such rural occupations. But why the hell not – if you like it, that's all that matters. I never knew you to take a decision you'd regret!

It did bother me, however, when my friend Priboye (the bearer of this letter) told me you said you wanted a quiet retirement and didn't wish to be bothered by urban types. Now I hope that doesn't include *me!* For one thing, if there's anyone today who claims my respect and admiration, that person is you – so how could I not be in touch with you? Besides, you brought me into that godforsaken newspaper after I left university and monitored my growth as a journalist until I rose to the rank of editor. Remember what you said on my first day at work – that I'd find real journalism was different from the stuff I'd been taught at school? But for you I'd have been fired on no less than four occasions for speaking my mind. If you've taken an interest in my affairs for so long, how can you now stop just because a gang of bastards took our jobs from us unjustly?

Priboye even tells me you refuse to read any newspapers. I know the papers usually get to your part of the country at least two days late. But let's face it, you gave more than thirty years of your life to journalism, so despite what's happened it's going to remain a part of you for as long as you live, whether you like it or not. Besides, so many things are happening today that are threatening to lead this country to an irrevocable explosion, that we cannot pretend to be unaffected by them wherever we may be. If you think – and I sincerely hope you do – that you have room in your retirement to consider a crisis that's brewing right by your doorstep, then please listen to me while I outline the main subject of this letter.

I'm going to try to be very brief, because I've no way of knowing if you'll be interested in the project. The point, simply, is this. You know very well how badly the traditional economy of the Delta communities has been faring as a result of two modern industrial projects which purport to enhance the economy of this country. First there is the Kwarafa Dam, which has severely reduced the volume of water flowing down the Niger and so curtailed the fishing activity in the Delta – and our people are nothing if not fishermen. Secondly, the spillage of crude petroleum from the oil rigs down there – one of which is in fact located near your own village – has proved an absolute menace to agricultural life, for many farms are practically buried in thick layers of crude, which kills off many fishes and other forms of life.

The latter problem has received a great deal of attention in the press lately, especially in respect of the resentment shown by the farmers and fishermen and their protest marches to the oil rigs. Not very much seems to be known about the Kwarafa Dam issue, and I'm quite certain the local fishermen themselves haven't heard a damn thing about it. But there is a radical group here in Lagos called the Committee of Concerned Citizens – CCC, for short – who have taken up the cause of the fishermen. Their general gripe is that the dam, which was meant to generate hydroelectric power and so support the take-off of much of the industrial activity of this country, has proved to be more a curse than a blessing. They cite the frequent power cuts stemming from the dam and have issued a few statements to the press on the woes of electrical failure. But in more recent months they appear to have devoted their energies to the goings-on in the Delta, urging that the dam should be closed – or, more appropriately, opened – to release water for the River Niger and so relieve the suffering of numerous populations along its banks, but especially in the Delta. They also condemn in very strong terms the destructive effects of the oil operations in the Delta country. You should see the language of their releases!

The brain behind the CCC's agitations over the Delta is its publicity secretary, a young man called Noble Ebika Harrison – nicknamed "Bickerbug" for his tirelessness and his no-nonsense style of approach as well as his rabid rhetoric – a veritable madman from Angiama whom I knew at Nsukka. You probably know him too, from his association with Isaac Boro in the struggle to liberate the Niger Delta from the Biafrans during the Civil War. Indeed you must remember him as the angry young man – a heavy-bearded mulatto – who frequently stormed into the *Chronicle's* offices during the war with denunciations of Biafra and even of the Federal army for its slow pace of liberation. Well, that's him. He teaches in a secondary school here in Lagos, and I can see his imprint behind every release the CCC has made so far. The Supreme Military Council, I understand, are watching him very closely. In fact it's rumoured that the National Security Service (NSS) has infiltrated the CCC with plainclothesmen hoping to break it up or maybe gain enough information about their intentions to nail the key figures of the movement.

What's all this got to do with you? Well, I doubt you'll deny that certain things are happening now which are bound to affect the history of our people in the Delta if not the whole country. I know how much your retirement means to you. But despite the forced retirement you are a journalist of the finest calibre who cannot shut your eyes to the events now unfolding before us. I believe it's time for us to rouse our investigative skills once again and produce a document that will be unrivalled in authority. You want nothing more to do with newspapers, I know, and I'm thoroughly in sympathy that you should avoid a profession that's betrayed you. What I propose is that we should simply exchange reports on events in the two sectors which are clearly related to one another. You will monitor the home front, as regards the growing resentment of the farmers and fishermen, while I cover the corridors of policy in Lagos and the agitation of the CCC (I've made contact with Bickerbug, in fact). Between us we should be able to follow the events to their logical conclusion and eventually produce a book that will long remain an authoritative testimony to the plight of our people, the Beniotu people, in these times.

You may wonder what's the point in all this Beniotu nationalism. Well, the events happening in Nigeria today have done little else but force us back to our little ethnic cocoons. And it is unfortunate that, as very often in these things, it is a few jaundiced individuals who rouse so much adverse ethnic awareness. If you doubt me, just look at you and me today. Is it any accident that we were the only two people on the *National Chronicle* to be summarily retired – and we are both Beniotu? As usual, no reasons were offered – the exercise was conducted by only one or two people appointed by the devil knows whom to compile a list of "redundant" personnel! I mean, how could anyone consider *you* redundant – or me for that matter?

It may well be that Murtala Muhammad meant well by his crusade against redundancy. I don't doubt it, but his intentions have been exploited by a handful of criminals who have simply seized the opportunity to settle personal scores. I believe some of our friends on the *Chronicle* must have felt uncomfortable that two Beniotu people – two *minority* people! – had risen to two of the topmost positions in the newspaper (one the editor-in-chief and the other the editor), without stopping for one moment to consider that we could have earned our elevation by dint of hard work alone and without the usual benefit of ethnic patronage. I don't know who exactly was responsible for our fate – there's been all kinds of speculation – but I very much suspect that bastard Ajibade, Chairman of the Board, who never tried to hide his bias against minority personnel, and who I understand enjoys the confidence of the military boys. I keep hoping that one day I will get my fingers round his throat!

Anyhow, let me stop there. Please let me know if this project appeals to you. I could think of no better collaborator than you, not simply because you are a compatriot (I abhor the word "tribesman", in spite of myself) but especially because I have always trusted your judgment. We must prove to idiots like Ajibade that we have something in us they cannot destroy or "retire". And if this book is the only thing we can contribute to justify the cause of the Beniotu people, I think it will have been a significant contribution.

I will normally write you through my friend Priboye Oruama who goes to the Delta frequently on some business transactions. I hope you will give your consent to this project and that in Priboye we may have a safe conduit for the rather delicate information of our mails.

My regards to Madam and the children – especially Boboango, who must miss Lagos badly!

Sincerely,
Piriye Dukumo

Brisibe Compound
Seiama.

5 December, 1976

Dear Mr Dukumo:

I got your letter of 25 August several weeks ago from Mr Priboye Oruama. Thank you very much for its contents. I have every respect for your feelings as conveyed in your proposal. But permit me to give my frank response to the issues raised, without any prejudice to your good intentions.

Priboye is right about my desire to spend my retirement, forced though it may be, in absolute peace and quiet. I know, of course, that there are often constraints on the wishes of men. For instance, the oil rig at Brikama is within relative earshot. The noise from the exploration machines reduces my desired peace somewhat. But I am thankful to the good Lord for the little peace He has given me and my family. And I mean to protect this peace jealously from the harassment of the wider Nigerian society, urban or otherwise. Yes, I do not entertain newspapers here: not so much because I want to turn my back on the news, as because I have had enough of the bickering and partisanship that has been the bane of Nigerian journalism and of our society.

I agree with you on the ethnic background to our retirement from our cherished profession, the "fourth estate of the realm". Ethnicity has become the major tragedy in the Nigerian body politic, and has hindered many a fine relationship among Nigerians. It is true I have desired to embrace the peace of my small rural world here, and to withdraw from the frustrations of national politics. But no one who has any love for our country will fail to lament the harm that ethnic sentiments, whether from Mr Ajibade or from anyone else, have done to her.

That, my dear Mr Dukumo, brings me to your project. I must confess that, despite my withdrawal from journalism, the prospect of reactivating my investigative instincts does tickle in me a fancy bred by over thirty years in the profession. But I am a little worried at the stridently ethnic character of your project. It seems to me a noble enough undertaking to investigate the disturbing events which you have identified. But why must the results of such an enquiry be directed towards justifying an ethnic group? If you indeed believe that the events happening now (to quote from your letter) 'are bound to affect the history of ... the entire country', why may your investigations not be geared towards exposing a malaise that is steadily destroying our beloved country, rather than towards swelling the ethnic pride of the Beniotu people?

Excuse me if I seem to be speaking contrary to your expectation. I am aware that a man like me who has chosen to withdraw from national issues and embrace the security of his narrow village world is a ready candidate for an ethnic project such as you have proposed. But I have seen enough of the woes of this country. Though I now enjoy the peace and security of my little Seiama, I have had the opportunity of living long enough outside of it to see things in a different light. I have nothing against the investigative basis of your project. But I must say that I find the ethnic chauvinism of it somewhat contrary to the caution which I have gained in thirty of my fifty-seven years.

Assuming we write such a book, what title do we give it – *The Woes of the Beniotu Nation*? My dear Mr Dukumo, I am a Beniotu man, and I love it. But I see no urgency now to advertise that tag when I know how much damage such tags have done in a multi-ethnic society like ours. While, therefore, I commend your journalistic insight in recognising a significant story, and the originality of the two-pronged approach that you propose, you must excuse me if I dissociate myself from a project which seems to me to have the potential of fanning dangerous sentiments. Besides, I am not getting any younger. A project like this calls for the energies of younger men like you. I have retired to this quiet life and, while I lament the meagre reward which my fellow fishermen and I derive from the craft, I am content to be one of their little company.

My family are well, thank God. Boboango missed Lagos for a while. But he seems to have fallen in love with Warri, where he attends Urhobo College. My wife joins me in extending our good wishes to your wife. It is our hope that the good Lord has in His infinite mercy brought some peace to your relationship.

Yours sincerely,
Tonwe Brisibe.

Lagos, 14.2.77

Dear Tonwe,

For Christ's sake please drop that formal tone of yours – "Mr Dukumo"! I know the matter of our correspondence is a serious one, but there's absolutely no reason for you to adopt such a stern approach to it nor to remind me how much older you are than I am. At thirty-seven years of age I don't think I'm such a small boy. And my wife is all right, thank you!

Now to the more serious business on which I invited your collaboration. I am surprised that a man like you who was driven by godforsaken tribalists in high places to seek refuge in your little community in the creeks should pretend to be supporting the cause of ONE NIGERIA on your battered shoulders. I am stunned that a man of your wisdom does not see the present trend of affairs in the country, that whatever remained of the ostensibly noble ideal for which we fought a civil war has steadily succumbed to the growing tide of parochial interests.

This country was split into twelve states in '67; barely two years ago it was split again into nineteen states – and what has been the effect of this? Before the division into states, people in this country had a larger sense of community in the broad multi-ethnic regional structures, and even a sense of national unity under a big federal umbrella. Most of us went to Lagos to make our living – you and I, for instance. But now the country has been split into little bits. The state governments are doing their damnedest to see that they get as large a share as possible of the "national cake"; even Lagos is trying its bit to convince Nigerians to promote their states, on the argument that the overall development of the country is best pursued through the growth of its constituent states. So what do you think is the meaning of all this? My dear Tonwe, we are wasting our time if we pursue the mirage of national unity when our own local people are living in squalor and disgrace, when big machines operated from Lagos are gradually destroying the resources that have traditionally nurtured our people since time immemorial. Who cares for national unity?

As to what title I would give the book, I really hadn't come to that point, but since you raise the question we might as well tackle it. I can read a certain contempt in the title you imagine I would give the book, but what the hell! What would you call the book yourself – *How to Build a Nigerian Nation*? The concern which your letter shows for the country at large leads me to believe this is the kind of title that would appeal to you. I must say that, despite my obsession with the problems of our people in the Delta – my "Beniotu nationalism", as you call it – such a title does have some possibilities. But who wants to be mounting the roof before he has had a chance to lay the foundations? And why should I give a damn about other ethnic groups when some of them have thought nothing of victimising my own if they see it fit and profitable to do so?

I respect you, Tonwe – honestly, you know I really do. Your judgment, your forthrightness, your dedication to ideals and principles – these are qualities I have admired in you all the years I have known you professionally. But it bothers me that you fail to recognise that Nigeria is becoming an increasingly ethnic society and that there's not a damn thing anybody can do about it. I don't know whose fault it is and I don't really care a shit. All I know is that nothing matters to me now more than the salvation of our homeland and the preservation of our heritage. Something has to be done to expose the insensitivity of these bastards here in Lagos and to prevent them from doing more harm than they have already done. I should expect a man like you, who has shown a lot of wisdom in going home to his roots and is even closer than I am to the problem I'm speaking about, to feel even more enthusiastic about the project than I do. Instead you sit comfortably in the "peace and security" of your "little Seiama" and tell me of your "beloved country" – so beloved it retired you in disgrace after more than thirty years of spotless service! As I said, I respect your wisdom and trust your judgment, which is why I approached you for collaboration in the first place. But, frankly, if you are not interested I am quite prepared to make other approaches.

Let me hear from you soonest.

Piriye Dukumo

Lagos, 10.7.77

Dear Tonwe,

Your long silence is eloquent. Priboye was in Lagos last week and confirmed that he personally delivered my last letter to you over three months ago. I would like to feel that you consider my proposal serious and weighty enough to be pondered long and hard. But something tells me not to be too optimistic...

All right. I'm prepared to admit that the tone of my letter was a little temperamental. It is not easy to be well-composed when one is discussing a matter that goes deep to the gut. But I admit that some of my lines may have caused you offence. For this I must apologise. You will of course recall that you and I had a most agreeable relationship all the years we spent together on the *Chronicle*. I know there were a few occasions on which you reprimanded me for my headstrong temper or what I remember you calling, in your characteristically accommodating way, my "over-anxiety". But you are the last person I would ever wish to offend, and I am deeply sorry if my letter gave you any offence.

However, my commitment to the project stands and I am still interested in your collaboration. Perhaps the "ethnic chauvinism" of my proposal is too strong. Well, I am interested in seeing whatever slant you think would make the project more acceptable. Never having written a book before, it is possible I have misunderstood how one is organised and indeed presumed on the patience of publishers for the kind of material I have in mind. If you can by any chance be persuaded to collaborate with me on this project – and I sincerely hope you can – then you may wish to suggest any relevant amendments. Nothing would please me (I almost said suit my ego) more than to have an acknowledged name like yours beside mine on the cover of the book.

My regards to your family, and thank you for your kind prayers for the relationship between my wife Tonye and me. I regret to say that she has refused to see reason and give me my due respect as master of the house. She prefers to listen to the counsel of her "liberated" friends, forgetting that though she is Beniotu like myself I will not hesitate to terminate this union (or whatever is left of it) if I feel I can't suffer it any more. Anyhow, more on that later.

I sincerely hope you will give my project a positive consideration, and I look forward eagerly to your reply.

Cordially,
Piriye Dukumo

Brisibe Compound
Seiama.

12 August, 1977

Dear Piriye:

You were not far wrong in surmising that your letter of 14 February put me off a little. I hardly see anything in my letter that could have merited such a severe indictment. But I thank you earnestly for the conciliatory tone of your last letter. I assure you that the goodwill which prevailed between us in our days at the *Chronicle* has not been shaken by this slight divergence of views.

It may come as a surprise to you when I say that I feel better disposed to your project now than I did at first. Something happened here recently which has shaken me from my complacency and inspired me with a concern not much below the zeal that you require of me. I have indeed become a little sadder than I was when I came here. I also believe that some record needs to be kept of the unfolding events, if only for the sake of truth.

On the 6th of this month, at approximately 4 o'clock, I was sitting in my compound mending one of my fishing nets, when a group of fishermen from Ebrima, a little community close to ours, approached me with sullen faces. I vaguely recognised one or two of them, but I made them all welcome nonetheless. I offered them our traditional refreshment of local gin. Before long, the leader of this team, an Aboh man by the name of Opene who has lived in these parts nearly all his life, put their business before me in plain terms.

Briefly, the group had gone to the oil rig in their village, owned by Atlantic Fuels, to complain that the enormous search-lights which they train on the waters around their offshore rig were drawing the fish away. They wondered if there was anything the company could do to save them the trouble; their lives depended on fishing, and they faced certain disaster if the schools of fish were forced permanently out of their areas of activity. The spillages from rigs and pipelines had done enough harm to their trade, and the activity of this new rig would only snuff out their lives for good. They were making this appeal to the company to see if there was any meaningful compromise they could reach with them. Apparently these search-lights were trained on the waters from dawn to dusk, and the delegation was asking if the lights could be switched off during those hours of daylight when the fishermen were engaged in fishing.

The delegation had sought audience with the white engineer. According to Opene, they had gone in a canoe to the off-shore rig, but were not allowed to moor the canoe so as to get onto the rig. The whiteman simply sat way off from them, cross-legged and pipe in mouth, while the mission of the group as well as the whiteman's response were relayed through an interpreter, a young man from Kpakiama by the name of Tebiowei, who according to my sources is employed there as Drilling Foreman.

The reply of the white engineer was uncompromising. He was fed up, he said, with these agitations and representations from illiterate natives who knew nothing about what the oil industry was trying to do for them. What, he asked, did the inconvenience suffered by a few scruffy fishermen matter to the general prosperity which oil had brought to Nigeria? He had no business with them, he said. If they had any complaints they should address them to their government, who had given his company the franchise to explore oil here. They were businessmen, and had no interest in political or social welfare, unless there was a visible need to pay compensation to families or communities whose land had been appropriated for exploration purposes. As it was, their business around Ebrima was strictly offshore and he didn't see any reason for them to curtail their operations just because a bunch of ignorant fishermen now caught fewer fishes than they were accustomed to. He said he hoped he would never see them again at the rig.

The delegation was naturally displeased by this arrogant treatment. A heated argument ensued. Not with the whiteman, of course – he had withdrawn from the scene – but with the interpreter Tebiowei. No doubt he hoped to impress his boss with how well he was protecting the interests of the company. And you can well imagine the discomfort of these fishermen who, despite their skill at manoeuvering the canoe, had to maintain a steady float on the water and conduct an impassioned argument at the same time.

But something even more disturbing happened at this point. It would seem that, while the fishermen were arguing with Tebiowei, the whiteman radioed the commander of a division of the Navy stationed in Warri. This division, established since the Civil War, is charged specifically with ensuring the security of oil explorations in the Niger Delta, and has units in Burutu and Forcados, near us. It has been rumoured for some time now that some officers and soldiers of these units have colluded with the oil companies around here in measures of dubious purpose directed against communities that have had cause to raise a voice in protest.

Anyway, while the delegation was engaged in confrontation with Tebiowei, a speedboat of armed soldiers taxied in at the rig and held float right beside the fishermen's canoe, almost knocking its poor crew overboard. Immediately, the dozen or so soldiers sprang at the ready, their guns cocked for action. Their officer wanted to know who was the leader of the "gang". The frightened fishermen turned reluctant faces towards Opene and muttered a few incoherent excuses. But Opene motioned them to be calm. He told the officer that the delegation was his idea, that he and his fellows meant no harm.

He had scarcely finished speaking when the officer slapped him hard across the mouth and sent him crashing across the bows of the canoe, dragging a couple of his men down with him. As his destabilised friends sought to steady him – he barely had time to struggle to his feet and clean the blood trickling from his mouth – the officer screamed an order, and the soldiers sprayed the waters close to the canoe with a welter of bullets. As they struggled to escape, these men who all their lives have mastered the buffets of wind and tide had trouble staying their craft on a steady course. Opene said that when from a distance he cast one final glance back at the rig, he could vaguely discern the officer shaking hands with the whiteman and receiving from him what looked like a package.

My dear Piriye, his words cut deep into my heart. You know how much more painful it is for a man to struggle to hold the tears back in his eyes than to actually shed them. And what do you tell a man who, though he has been charged to represent in his report the collective threat to his community's livelihood, affects you even more deeply by the picture which he presents of an injured pride? For a while I could not say a word in response. But it was becoming increasingly clear to me what was being laid on my shoulders. I have come home to lead a peaceful life, but how can I honestly disown a cause which has everything to do with the peace that I crave?

'Well,' I said, 'what do you think I can do?'

'We have complained and complained,' said Opene, 'but nobody seems to listen to us, because we do not know book. You know as much of it as they do, and they will have to listen to you. We believe we have no course left to us but to take our case to the Military Governor at Benin. Mr Brisibe, you are one of us now. We think you should speak for us. Our people are suffering, and will be wiped out unless something is done at once.'

'What about the influential men around here, like Chief Zuokumor and the council of elders?'

'There is nobody we have not been to,' Opene said.

I had mentioned Councillor Zuokumor only as a matter of form. But everyone here knows the corrupt, ignominious roles he played in two recent arbitrations: one in the dispute over fishing boundaries between the villages of Dimiso and Erefiebi, the other in the matter of compensations to be paid to the village of Ekpetiama by the Dutch company, Atlantic Fuels. In all sincerity, I would not seek his intervention in any matter, nor did I really want to encourage Opene and his men to do so.

I have made plans to go to Warri in the very near future, to seek an audience with the authorities. Upon my return I will keep you posted on the outcome of my mission.

My dear Piriye, I have gone into so much detail about the fishermen's experience, not only to give you a vivid picture of what is happening around us here, but especially to highlight one aspect of it which relates closely to the investigative project you have proposed. I was struck quite deeply by the personality and attitude of Opene. That an Aboh man should show such sincere commitment to a cause which affects an essentially Beniotu community, makes me think deeply. You should have heard how often and how passionately he used the phrase "our people"!

I do not wish to belabour the ethnic issue which I raised in my first letter to you, especially since in your most recent letter you seem to have retracted somewhat from your original conception of the project. But I simply want to provide one more reason why I think we should see our goals in a much more human or perhaps national light than an ethnic one. For if the prospect of inter-state or inter-ethnic conflict is as dismal as you presented it in an earlier letter, would this "outsider" not be relegated to the background by his Beniotu colleagues, rather than be trusted to lead them in a cause which will determine their ultimate survival? You may say that, having been sheltered by the Beniotu for so long, he has no choice but to consider himself one of them. But you must know that, if indeed we are doomed to fragmentation, there are plenty of troublemakers around us who will encourage discord by playing on the ethnic difference between Opene and his fellows at Ebrima. Only time will prove the point.

No doubt the foregoing paragraphs will have shown that the project which you propose has touched deep chords in me, and roused the professional curiosity which the forced retirement of the last few years may seem to have laid to rest. Yes, I will collaborate with you on it, if you will consider kindly my advice that we beware the limitations of the ethnocentric approach. Perhaps we should also suspend for the moment the question of what title to give the book, or even the chapters around which it should be conceived. I have had no more experience than you in writing a book. My journalistic instincts also tell me that we should concentrate meanwhile on an exchange of investigative reports. When the time comes, we may see clearly enough the lines along which to order our material into a book. A little patience will help here.

About the Committee of Concerned Citizens: I do indeed recall the Harrison whom you mentioned in your first letter. There is something rather threatening in his personality, and I am not surprised to hear he is involved in yet another cause. Let us hope his role is a salutary one. I expect to hear more from you on him and the Committee.

I have heard what you said about your problems with your wife. Please do not take a hasty course of action over the matter. Try to make love and understanding the sole guides to your actions. In all events, please consider first the chances of reconciliation rather than the break-up which you threaten. I say so not because you are both Beniotu but because, like two good human beings, you should recall it was love that brought you together in the first place. My wife joins me in my plea.

Best wishes.

Yours sincerely,
Tonwe Brisibe.

Lagos, 30.8.77

Dearest Tonwe,

You can't imagine how overjoyed I was to receive your last letter. Yes, yes, yes – I heartily welcome your advice that we put aside for the moment the question of titles and chapters and concentrate on an exchange of investigative reports. You're damn right. The book will come – *it will come.* In conceiving it, I simply couldn't resist seeing us in the image of Woodward and Bernstein – the *Washington Post* team that produced *All the President's Men*, you remember? – although you have left the profession and a considerable distance separates us. But I accept your advice wholeheartedly.

I was also moved by your report on the events at Ebrima and your growing involvement in the situation in the Niger Delta. Since I last wrote, a few interesting things have indeed happened here in Lagos. Bickerbug Harrison and the CCC have intensified their activity, and they seem to be operating on about three fronts. The first of these is official. Some of the members of the Committee have succeeded in establishing a beachhead with the Supreme Military Council here in Lagos, and have been put in direct touch with the relevant officials at the Ministry of Petroleum and Power.

The second line of operation by the Committee connects them with the Nigerian press. I don't know how much of this you know, especially since you have decided to have very little to do with newspapers these days. But nearly every week now there is something or other in the papers about the deteriorating situation in the Delta. These are articles – some of them by members of the CCC, others commissioned by them, yet others by various interested writers – denouncing the role of the Kwarafa Dam and the oil industry in the *Chronicle* – naturally, as it is the paper with the widest circulation, although I am a little surprised to find that they are given so much and so frequent space to attack the government, considering the government has majority shares in the newspaper.

Incidentally, I have in the last couple of weeks reestablished some contact with the *Chronicle*, in view of their close links with the Committee. I don't often go there, of course – not after all that's happened to us. My contact there is Miss Latifat Ogedengbe –

you remember that dark, slim girl, always with a smile on her face, a graduate of Mass Comm. from the University of Lagos, who worked for me as a reporter on the arms smuggling beat? Well, she's the one. She's been very kind and forthcoming on the relationship between the CCC and the *Chronicle* – so gracious indeed that despite my ethnic reservations I find myself impressed by her personality as well as beholden to her. It would seem that the newspaper has done some censoring of the material from the CCC. Lati has been kind enough to let me see some of the articles, complete with classified information and some very incriminating photographs of the damages suffered by Niger Delta communities, which the newspaper has refrained from publishing – at least in the form in which they came. It all looks very disturbing indeed, and every day that passes makes me an even angrier man than I may have seemed so far in my letters to you.

The third and potentially most dangerous line of action taken by the CCC is the mostly one-man crusade of Bickerbug himself – on the soap-box, etc. About two weeks ago he (or more correctly the Committee) placed an advertisement in the *Chronicle* for a public address he would make at Tafawa Balewa Square under the title 'The Woes of the Niger Delta'. But he was not allowed to make that speech. A day or so before the event he was visited by some plainclothesmen and warned not to show his face at the Square. He took the warning, but was determined to address himself to the public nonetheless. He therefore decided the best thing to do was an impromptu, unannounced affair at Campos Square, in the very heart of the city. He made it. A week ago he stood on a platform there and heckled passers-by – Campos Square is, as you know, well known for such scenes, so he had no trouble attracting an audience in a very short time. He was there with one of his colleagues on the Committee – Tari Strongface. He spoke fearlessly about the insensitivity of the government, the corruption of the officials, and how all this was leading to the suffering of the masses. I don't think I ever heard such strong language used by anyone in all my life. Fantastic! He has such firm command of the language. Even more impressive is his command of the ideological idiom. Not that he was throwing around the fashionable names and jargon of Marxist-socialist ideology – that wouldn't work with the sort of crowd you find in Lagos, often idle and not very literate people. Bickerbug simply chose his words quite well and besprinkled his speech with appropriate cases of dehumanisation. I must confess I was never much into this ideological stuff, but after listening to him and reflecting on the connections between the ideas he explored and the situation here today, I certainly think it all makes a great deal of sense.

Bickerbug had begun to show pictures of the devastation done to the Niger Delta environment by the oil exploration – deforestation of on-shore sites, desecration of traditional shrines, evacuation and tearing up of whole villages and farmlands, vast areas of oil spillage and great quantities of aquatic life destroyed in the wake, etc., etc. – when the police raided the scene and pounced on him. There was a mild scuffle. Not with Bickerbug – for although he poured defiance on the police he never resisted arrest – but with a handful of the listeners who by now showed a little more than sympathy for the message of their speaker. Anyway, the police grabbed Bickerbug and about three or four of the resisters and threw them into the anti-riot van they had come in. They spent two nights at the Panti Street Police Station before they were released.

How do I know all this? Well, I was there at the Campos Square event. In recent times I have reestablished contact with Bickerbug – that is, since our dealings with him at the *Chronicle* in his activism with Isaac Boro during the Civil War – and have come to know him quite closely. Getting acquainted was not easy, because he was very guarded at the start. When I found out his residence at Okesuna Street and paid him a sudden visit about ten days ago, he was not immediately hospitable. That didn't surprise me at all. Activists, especially in these troubled times, have a way of taking you to be a member of the establishment until you convince them otherwise. The first statement that came out of his mouth when he opened his door to me – eyes wide and flashing – was, 'Who told you I live here?'

I felt rather awkward. But I smiled in spite of myself and told him he was not so well-guarded as he thought he was – anyway, I was interested in his message and had asked a few mutual friends to show me to him. He looked me over before showing me reluctantly to a seat.

'I hear you were retired from the *Chronicle*,' he said, after an uneasy silence.

'That's history now. Who told you?' I thought I should give it back to him – what the hell!

'Was it supposed to be a secret?'

'Not really. But I didn't think you were interested in such things.'

I think he got the message, because he flashed me a look from the corner of his eye. I thought the atmosphere of hostility would not help my mission very much. So I tried to be friendly.

'Well, how are things going?'

'What things?' he asked.

'You know. The Committee, the crusade. You know.'

'So far so good.'

'I'm sure there are problems. I was over at Balewa Square the other day for the rally you'd advertised – a number of other people were there too. But we waited in vain for your appearance.'

'Yes, I know,' he said, lying back on his bed. 'Didn't you also hear they stopped me?'

'Well, I guessed there was a problem. Was it the police, or Security?'

He breathed deeply and took his time to answer.

'Security. Let's talk about you for a minute, sir. Why do you suppose you were retired from the *Chronicle*?'

'Well – you know, er,' I hedged. Honestly, I was taken aback, but being in his residence I supposed I was at his mercy. 'Well, you know, the usual Nigerian problem – prejudice, vindictiveness, high-handedness. You do the best you can at a job, only to find you are hurting someone else's image. Or somebody doesn't like your tribe – *especially* if somebody doesn't like your tribe.'

When I looked up at him he was nodding in agreement.

'I don't know if you know this, but Tonwe Brisibe and I were the only ones retired from the newspaper. I mean, even if anyone had any objections to my person, I don't see how they could have felt the same way about Brisibe – he is the nicest man in the whole world. That's why I think they played a foul ethnic game with us.'

He was listening with rapt interest.

'Who would have suspected you are Beniotu?'

'Why do you say that?' That certainly hurt, and my brow screwed up a little.

'Well, I sometimes wonder – you Beniotu who have lived so long in Lagos and were probably born here – we sometimes wonder if you are true Beniotu or just coastal Beniotu and only visiting home when...'

'I'm *not* coastal Beniotu and I wasn't born in Lagos,' I cut him short. 'I grew up right in the Delta and I know my roots as far back as they go.' I thought the idiot was pushing me too far. I couldn't stand it.

'Okay, okay. No sweat. I just wanted to be certain who I was talking to.'

He seemed reassured, for a little smile played on his bushy face and he raised himself a bit from the bed.

'So what do you do these days?' he asked.

'Oh, this and that,' I said, 'but mostly freelancing. I use different pen-names for different papers. But I never write for the *Chronicle*. I think I've had enough of them.'

He was looking steadily at me and nodding a little, more I think for fellowship than in agreement.

'Right now,' I continued, 'I want to follow what's going on with you and the Committee because I think some history is about to be made and may indeed be happening already.'

'History!' he spat, getting up and walking towards the window. 'Look, man, I've had enough of history. When I worked with Isaac Boro during the Civil War I had illusions about making history, a romantic sense of mission. But it all came to nothing. He was killed under questionable circumstances, largely because a fight which should have been taken care of by bona fide Beniotu citizens got bungled by strangers who had no stake in the matter.'

'Strangers?' I wondered.

He looked me straight in the eye, then shook his head and got back on the bed.

'Look, Piriye,' he said, lying on his back again, his head against the wall. 'There are two things I don't do very well. I don't know how to fool around with too many words when I can say what I want to say in a few direct words. I also don't ... I also don't ...' he seemed to hesitate, looking me over once again. 'I don't function very well in a group whose purposes are rather diffuse.'

Something dawned on me as I nodded at his words.

'You mean the Committee?'

He nodded. Those steady eyes, made even more fearsome by

the bush of hair on his face, were burning a hole through my brain. Bickerbug's advertisements of his public address – the two that I've seen so far – are supposed to be a joint Committee affair, but none of them could be said to have been truly released by the Committee. As far as I could recall, the text of the release in each case gave full prominence to the personality of the speaker and it was clear he was the prime mover in the affair. And, come to think of it, in the speeches made so far by Bickerbug, I didn't think I had seen more than one or two members of the Committee alongside him on the platform.

'Are you saying you don't think your colleagues are sufficiently committed to the crusade you are waging over the problems in the Niger Delta?'

'No,' he said, 'not really. But tell me, could you honestly expect them to feel the same level of commitment as those who are directly affected by these problems? I know we all talk broadly and glibly about the problems in the country. But sometimes these general arguments stick in my throat because I believe in confronting a problem by practical means.'

It made little sense to me. I sighed and shook my head in bewilderment. I wanted to ask him a few more questions about his colleagues – their backgrounds, their real stakes, even their roles. But he seemed to be in a hurry to go somewhere, and cut me short.

'Piriye,' he said, 'I know you are a journalist, even though your friends in the profession played you foul, and I'm always very careful how I talk to you people. But I also know you are a Beniotu man like myself. Nobody is going to do anything about the suffering of our people if we don't do it ourselves. Let me tell you something,' and he drew closer. 'Don't think I'm going to spend my life making speeches on the soap-box, and holding useless council with the Committee. When the time comes that I feel my friends are not sufficiently behind me in what I'm trying to do, I'm going to cut loose from them. I have a mission, and that is the salvation of our people from the inhuman devastation which policies formulated by corrupt officials here in Lagos have constantly brought about. The fight which Boro started continues, as far as I'm concerned, and I'm prepared to give my life to this mission if need be, just as he never hesitated to commit his own life to the struggle to liberate our people during the war. You'll find out more as time goes by.'

He grabbed my hand in a conspiratorial grip, and from the way I gripped back he could probably tell, even without my saying a word, that I did not consider myself alien to his sentiments. Let's face it, Tonwe, the man had already got me to acknowledge my Beniotu nationalism, to swear I was a true son of the tribe – so what could I lose now by expressing my solidarity with a pro-Beniotu cause, even if tacitly? I know what your reservations might be: 'How can you hope to conduct your investigations into this business with due professional objectivity when you have already started by taking sides?' I can already hear that question as you read these lines. But let me assure you that I am as committed to seeing us make the best of this project as I am ready to stand up and be counted with the race. That was exactly the feeling that ran through me as I shook Bickerbug's hand.

'Now your speech has been cancelled by the NSS, what's next?' I asked as I moved towards the door.

'The NSS hasn't cancelled my speech,' he said, with a self-confident smirk on his face. 'They have simply altered my plans for it. Come to Campos Square on Tuesday, at four p.m. I'm not announcing the speech beforehand, so those jokers won't get in my way again. But I know that as a good Beniotu man you can be trusted to keep a secret.'

That was the speech during which he was arrested and jailed for two days at Panti Street. I wanted to pay him a visit there, but I wasn't sure the police would let me. Besides, it seemed to me wiser to avoid being openly identified with him so soon. Not that I was any less sympathetic to his nationalist sentiments. But if I allowed myself to be so readily drawn into his activist circle and so fall into the net which the NSS have obviously thrown around him, I would be robbed of the opportunity of observing this unfolding drama in its entirety and documenting it from whatever perspective I chose.

Which brings me back to the point you have stressed in your two letters so far – that we avoid the path of ethnic sentiments in this matter. I want to assure you that I am considerably less chauvinistic than I was at the point when I wrote my first two letters. Your reassurances have been in a large sense responsible for the mellowing of my temper. Perhaps it was inevitable that, as I got into the business of investigating the events around me, the old professional sense of duty put the usual constraints on me. Or perhaps this is all part of my maturing process. Whatever the explanation may be, I want to assure you that I know what our main objectives are in this project and that I'll execute my side of it with the energy you have had cause to admire in my work.

Still, the slant which Bickerbug gives to his crusade clearly indicates that, however hard we try, we *cannot* evade the ethnic factor in our investigations – to the extent, that is, that we are prepared to recognise Bickerbug as a key factor in the unfolding history of the Delta. We may pretend to pursue this project with all the objectivity that the old profession enjoins on us. We may go on collecting our data without bothering at all under what sorts of headings they will fall later. We may even let ourselves be guided by your plea, that we see the troubles in the Delta as part of a larger human drama rather than as a matter of isolated ethnic goals. But, with all due respect for your professional wisdom, do you sincerely believe it is possible to keep the goals of our investigation in strict isolation from the motives of a man who – at least as far as I can see – has emerged as the principal factor in these events? I only hope I can be as careful and as discriminating as you are. But I can't trust myself.

My domestic life is deteriorating. But I don't want to bore you with banal details and end this serious letter in an anticlimax. Besides, this has been a long letter. I have taken the leisure of Priboye's visit to Lagos to write you in as much detail as possible so as to give you a clear picture of the development of things at this end. I can only hope your interest has been sustained. Let me therefore postpone my report on the situation between my wife and me for a future letter. All I can say now is that, unless something dramatic happens to change the course of events, I cannot see us continuing as man and wife for very long.

My regards to your family, and please write soon on the interesting developments at your end.

Cordially,
Piriye

Brisibe Compound
Seiama.

18 September, 1977

Dear Piriye:

Thank you for yours of 30 August. Before I go into the main business of our correspondence, let me take up the issue with which you ended your letter – that is, the sad state of your marriage.

Not being in Lagos any longer, I am in no position to give you direct assistance towards settling your differences. I can only implore you once again to exercise restraint and seek the path of accommodation. Consider also that there are a few of our people in Lagos who would be willing to mediate between you if you would put the case before them in the good old traditional way. I can think of Messrs Okudu, Enekoru, Egbuson, Tobi, even my cousin Long-John Ebiegberi. I know how delicate these marital problems are, and I can only guess at what may be in some way responsible for the decline. But whatever it is, please give good counsel a chance to prevail over wilfulness.

You are quite right in recognising Mr Harrison as a principal factor in the development of events connected with the life of the Delta. I am a little worried, though, about the course of his relations with the police and the NSS. I am against violent confrontation in any form and at any level. I have no doubt that he is engaged in a worthwhile cause: to the extent, that is, that he will move the authorities to pay greater attention to the plight of the people in this part of our dear country. But I doubt that he will achieve very much if he sets himself so decidedly on a collision course with the authorities, rather than explore the possibilities of mature and meaningful dialogue.

There was indeed some indication of this in your letter. You mentioned that some members of the Committee are charged with consulting with the Supreme Military Council and the Ministry of Petroleum and Power. I should suspect that their consultations have a better chance of bringing a meaningful change to the situation here than the dangerous approach of Mr Harrison. I was not surprised that he was warned against speaking by the NSS. He does not seem to realise that we are still in a military regime, and that the embargo on public meetings has not been lifted. While I have no wish to dissuade you from probing the involvement of Mr Harrison in this matter, please mind how closely you identify with him. Try to avoid identifying yourself with the sort of explosive measures he seems to be contemplating. He talks about Boro and about fighting and laying down his life. I see no future in this but one that spells disaster, not only for him but for the people he claims to be concerned about. I do not believe that our people are ready for the sort of violent revolution he preaches. So please tell me more about the consultative efforts of other members of the Committee.

As I pointed out in my last letter to you, I have been drawn willy-nilly into the events developing in the Delta. But I have made it quite clear to all the parties concerned that I will have nothing to do with any measures which do not appear to me to be aimed at achieving a peaceful solution to our problems. The moment I discover that my name or my efforts are being pressed into the service of violence, I will withdraw my participation forthwith, no matter what those who look up to me may think of my course of action.

Before the delegation of fishermen which visited me on 6 August left my place, I said to them that I was going to think over their appeal and let them know how I thought we should make our approach to the military authorities. I did think about the matter, long and hard. Eight days later, on the fourteenth (exactly two days after my last letter to you), I began by trying to acquaint myself a little better with the lives of these people who had come to me. I must admit I had all this time been a little selfish in my desire to enjoy a quiet retirement. In the last few weeks, it has become clear to me that this peace and quiet may elude me if I shut my eyes to the all too obvious suffering of people around me. So I decided to address the problem face to face, at least satisfy my humble conscience that I have played my part, however limited. So, having thought well over the matter, I made up my mind to pay a few calls. Naturally, the first man I visited was the leader of the delegation himself, the Aboh man Mr Opene.

My visit to this man's place confirmed my high regard of him, gave me a deeper sympathy for the honest, unpretentious life of our folk, of whom I now proudly count myself one. Now, I am no longer trying to canvas you into adopting a humane as against an ethnocentric approach to our investigative project, by painting a holy picture of a man who offers selfless service to a community where, at least in the context of the parochialist logic which dominates the minds of many of our countrymen today, he is an absolute stranger. I believe I have enough faith in your mature professional judgment not to want to put anything over you. But to cut a long story short, Mr Opene is a truly decent man. He is married to a woman from Kiagbodo. Between them, they have five children. They seem to me to live in absolute harmony. He is a man very highly respected in these parts, as I have come to realise. From the ease with which he speaks our language, you might think he had much deeper roots here. With regard to the problem in hand, I could see that he had a much greater awareness of its implications than you would credit to a simple rustic man like him. After we had exchanged the usual civilities, and he had gone ahead to entertain me in the good old traditional manner, we spoke about the oil problem.

'Mr Opene,' I said, 'this business with the oil company is no small matter. We should handle it very carefully.'

'I know,' he said, nodding. 'There is trouble in it.'

'You know what kind of trouble?' I asked.

'Oil is money,' he said, spreading his fingers as though preparing to count the costs. 'Money for the government. Money for many people. But not our people. And they do not mind what they do to us so long as they protect this money from troublesome people like us.' He smiled and shook his head.

'But you don't seem like a troublesome man to me, Mr Opene,' I assured him.

'I know, but does it matter to them?' he queried. There was a look of injured innocence in his eyes. 'How could anyone suspect poor fisherman like us of starting any trouble? All we went to that machine in the waters for was to make a simple complaint. But I could tell, by the force with which that officer struck me across the eyes, that the tone of our complaint was not the issue for him. His hand told me all I could ever want to know. That was when I realised how much trouble this oil means for our people.'

I wanted to know what else he knew.

'We are not the only people troubled by this oil matter, you know,' I suggested.

'I know,' he said, not raising his head or his voice. 'They have another one of these machines in the waters near Kuruma. The village has been completely wiped out by the floating oil, and most of the people have resettled in Burutu. There is another one near Birebe. There is practically no fishing life there any more, because the fishes die and float on the black surface of the water in large numbers. Many of the people have moved out to resettle in places like Emevor and Igbide. There are others. We hear about them all the time. Many of the people come this way from time to time, and we too go there sometimes, to see for ourselves. Mr Brisibe, we live in troubled times. I know that very well.'

'And has there been any trouble between those communities and the authorities, the kind you had here with the soldiers?'

'The Kuruma people protested at Burutu, and many of them were beaten and locked up for days in cells without food. One of them died in the cell. He was quietly buried. Nobody has heard any more about the matter. Mr Brisibe,' and he looked up at me, 'we know how much trouble we face. But is it not better for us to say something, in the hope that the situation may be different from the one at Kuruma, than for us to keep quiet and be wiped out in our silence? Tell me, for you surely know better than we do.'

I did not say anything. Not because I had nothing to say. But because I was struck by how little information I had about an environment in which I had chosen to spend the rest of my life. I was equally struck by the depth of concern shown by a man whose simplicity you would easily be inclined to equate with ignorance. I sighed, and nodded at him.

'Mr Opene,' I said, with a growing feeling of commitment, 'we will have to do something about this. You and I.'

He grunted his assent, and drew nearer, like one willing and eager to bear his share of a heavy load.

'We shall go to the Brigade Commander in Warri, you and I, and make our case as carefully as we can. He must have known what happened at Ebrima. I don't expect him to do anything about the soldiers who gave you trouble there. He may even have sent them himself. But let us build our case carefully. We shall not make any trouble. I want you to be clear about that, you understand?'

He nodded, adding, 'I am not a trouble maker, Mr Brisibe.'

'I know. I know,' I assured him. 'Neither am I. That is why I am suggesting that only you and I should go, rather than the entire group of fishermen, as in Ebrima. You know what soldiers are like. Once they see a crowd of men, they get in the mood for a fight. And we do not want that.'

That was on 14 August. I did not visit any of the other fishermen. I was so impressed with Mr Opene's personality that I was quite confident of his ability to represent the others as fully as possible. He visited me once more, about two days later. Last week, on 10 September, he and I made our visit to the naval officer, Commander Bayo Adetunji, in Warri. It was about noon.

You may not know it, but I have always been more careful in my relation with soldiers than with other people. I think I can safely say that I am a peace-loving man. But somehow, deep inside me, I feel like a slumbering volcano, for I am not certain that I can trust or control my conduct when I have been deeply, subliminally provoked. It has not happened many times in my life. In fact, I cannot recall any other occasion when I lost my calm than at that party in Ikoyi in 1972, when some overzealous Permanent Secretary threatened to deal with me in particular, and the *Chronicle* generally, because, as he charged, we were not giving the government's policies sufficient publicity in our pages. You remember, don't you? You were at that party yourself. With the military boys, I have been content to keep a studied, polite distance. When Mr Opene and I visited Commander Adetunji, it became clear to me why I have always felt that way.

To begin with, it took us a good while to get an audience with the officer, on the grounds that we did not have a previous appointment. But more than that, one soldier took special exception to Mr Opene's presence. He tried to order him out of the garrison. I tried to reason with him, as calmly as I could, saying that Mr Opene's presence meant everything to our mission. I suspect he may have participated in the operation against Mr Opene's party at Ebrima, and thought we had come to make trouble. Luckily, there were a few other soldiers around who treated us with a little more decency. We finally gained our audience.

Commander Adetunji received me pleasantly, even recalling old times.

'Hello, Mr Brisibe,' he beamed, stretching out his hand to me across the table. 'What a pleasure to see you here. We miss your excellent articles in the *Chronicle.*'

'I'm pleased to meet you too, Officer,' I said, returning the compliment.

'Do you live here in Warri?'

'No, I live in my village in the creeks, in Seiama.'

'You don't look like someone who would fit into village life,' he joked.

'I didn't think so myself, until I had to do it.'

There was a brief, courteous silence.

'Well, sir, what can we do for you?' he asked.

I was not happy that he never even looked in Mr Opene's direction, as if the man did not exist. I was, in fact, the one who asked Mr Opene to sit down.

'This gentleman here,' I motioned respectfully, 'is Mr Opene. He comes from Ebrima, a little fishing village close to mine.'

The officer nodded at me, still ignoring Mr Opene's presence. I continued, 'We have come here because we need your help. We fishermen in the creeks have no other source of survival than the fishes. Now, the activities of the oil companies around us are posing a threat to our survival. We have tried to complain to them, but I don't think they are prepared to listen to us. I would have thought ...'

'What exactly is your complaint, Mr Brisibe?' the officer cut in. I could see that his countenance had changed. The smile had completely disappeared. He was looking every bit the stern military man. I swallowed my embarrassment. He might have his reasons.

'A few weeks ago,' I said, very calmly, 'there was an incident at Ebrima.' He was no longer looking at me. He now engaged himself in sketching something or other on a pad in front of him. But I continued, 'Mr Opene here had gone with some of our colleagues to the off-shore rig of Atlantic Fuels, to register a request. The searchlights which the company directs on the waters have been disturbing the movements of the fish. All Mr Opene and his little group had gone to request was that the lights be turned off during those hours of the day when they are engaged in fishing. They made their plea as peacefully as possible. But apparently ...'

'Were you there with them, Mr Brisibe?' he cut in again, still sketching on his pad.

'No, but I can ...'

'Then how do you know they were peaceful?'

'These are humble fishermen, Officer. They would not ...'

'You're not answering my question. How do you know they behaved peacefully?'

I really could not answer the question. I had only Mr Opene's word on the conduct of his group at the rig. But I would be prepared to swear to the integrity of the man and his claims. The officer, however, took advantage of my momentary bewilderment to press his point.

'You see, Mr Brisibe, you've only just come to these parts. We know these fishermen and their tricks on the oil companies. It's either that the explorers have destroyed the shrine of their god, or that their fish traps have been damaged, or that the tides have been disturbed, and they can no longer catch fish when they normally do. Or some other silly complaint. They tell us anything just to get compensation. Compensation, compensation – everybody wants to be paid something. That's the trouble with this country today. Money, money, money.'

All this time Mr Opene was very quiet. Of course, he does not speak any English, so any communication between him and Commander Adetunji would have been through me. But the officer simply behaved as if Mr Opene was not there in the room.

'There was something else, Officer.' He still did not look me in the face, but there was now a wry smile around his mouth. 'While Mr Opene and his men were at the rig, a small company of soldiers ferried up to them and ...'

'I know what you're referring to. I sent those men myself. The exploration engineer radioed to tell me some fishermen were threatening trouble at their rig, so I sent my men to restore order there.'

'But, Officer, the way they handled those helpless, innocent men ...'

'Mr Brisibe, I think you are a gentleman, and I would like to continue to respect you. But please try to extend a little respect to those of us here who are trying to do our jobs. I know this is your home area. But I have been here long enough myself to know the people I am dealing with. You call these men helpless and innocent, and you seem to imply that my men are violent or ruthless. Well, we have been called all sorts of names before. But we have a duty to do here. And, please, I would appreciate it if you didn't try to teach us how to do our jobs.'

For a while after that, there was silence between us. It was becoming clear that I was getting nowhere with him. He was not prepared to reason or listen to me. He was not even looking at me, let alone Mr Opene.

'Well,' I finally said, 'what do you suggest we do?'

'Frankly, I don't know,' was all he replied.

After a while I rose. So did Mr Opene.

'Thank you, sir, for your attention,' I said, as I stretched my hand to him. 'I am sorry we bothered you.'

'Don't mention it,' he said, rising and shaking my hand. He smiled once more. It was clear he was glad we were leaving. I was boiling inside. I was so disturbed that, on my way out, I opened the door of a wall cabinet instead of the door of the office. He left me to my fumbling. Eventually, I found the right way out.

On our way home to the creeks, I was lost for words to say to Mr Opene. For one thing, the officer's attitude had erased that margin of doubt which, never having met him before, I was prepared to allow in his favour. Although I have suffered the extremes of bad faith in my retrenchment from the *Chronicle*, I still have that old tendency to believe that every human being has a natural aptitude to deal fairly with you if you treat him or her with due respect. Frankly, Piriye, I think Commander Adetunji shook my faith rather strongly. Throughout our return journey, I was left wondering what his real motivations were, and how far his links could be traced.

For another thing, I was a little ashamed that I had not fulfilled the expectations those fishermen had of me. In fact, I was beginning to wonder if I was properly qualified to represent them in their plight. Although, whether I like it or not, our fates are now inextricably bound together, I wonder if I am not bringing an undue complication to their lives. Before I came here, they bargained their demands with the authorities in their own peculiar style, and somehow or other they managed to gain certain compensations. At any rate, the authorities humoured their simplicities and traded off a few concessions to them. Now, here I come, an educated man from Lagos, thinking I know too much and trying to upset the pattern of relationships here. Tell me, Piriye, in your honest judgment, do you think I am the sort of person to lead them, when all I can ever bring them is trouble?

I am sorry if what I have said seems to indicate that I am trying to back out of my commitment to your project. I know I have my reservations about your original conception of the project. I am also no less cautious about some of the personalities whom I now see featuring in your investigations – I mean the likes of Mr Harrison. But I do not like what I see here either. I will continue to pray to the good Lord to ensure that we achieve the proper changes as peacefully as possible. But I cannot now turn my back on people who have seen fit to put their trust in me.

My plan is to go to Benin in the next week or so. I believe I still know a few people there who can afford to listen to me. Perhaps I will have a good chance of not only putting across the grievances of our people but also assessing what are the real odds working against us. Whatever is the outcome of my visit, I will let you know.

Your friend Priboye tells me he will be going back to Lagos by the end of this week. So I will be giving this letter to him tomorrow. What, really, is his business? I am not trying to pry into his life. I simply want to be sure who is carrying my letters. Since you seem to trust him, I am prepared to take your word.

Our regards, as ever, to your wife. Please remember what I said in the opening paragraph of this letter. God bless and keep you both.

Yours sincerely,
Tonwe Brisibe.

Lagos, 12.10.77

Dear Tonwe,

Priboye obviously left the Delta later than he had planned, because I got your letter from him only last week. But he's all right. You can trust him as much as I do. I frankly don't know his *real* line of business. All I know is, he's a trader – exporter, importer, that sort of stuff. But he's okay. I've known him since our boyhood days in Warri. And I can vouch for his integrity. He's a decent Beniotu man. And I like him a lot. Our business is safe in his hands. But I haven't told him anything about what we're doing.

I'll be damned – your letter made *heavy* reading! Things are really happening down in the Delta. I know you think I'm being hasty in conceiving of the headings under which the projected book will be written. But I can see from what your letters say so far that a whole chapter or section could gainfully be devoted to the theme of 'Private Profit' – whether for the oil companies, or for the petty soldiers who seem so committed to protecting the interests of the former for their own blasted purposes, or even for the local fishermen who don't mind bargaining for compensation now and then. And why the hell not? These bastards with their big machines and their big talk have been exploiting the wealth of our land all these years – what have our people got out of it? Nothing. And what are these paltry compensations compared to the loot the big ones make away with *every day*? Damn them.

Events are also gathering pace in Lagos. Bickerbug and I had an interesting session two days ago. It seems the split within the CCC is widening. If you hadn't parted company with newspapers you would have seen a report in the *New Nigerian Times* some two weeks ago about a meeting between certain members of the CCC and the Minister for Petroleum and Power. I'll first tell you the version from the newspaper, then I will tell you Bickerbug's interpretation of the Committee's moves, before ending up with my own assessment of the whole situation.

On the 24th of September some three members of the CCC – without Bickerbug, as you can imagine – went over on appointment to the minister to discuss the Delta problem. No journalists were allowed – it was all done behind closed doors – but an official communiqué was issued by the minister's press secretary, Tolu Adeoye, and this was carried by the *Times*. It read as follows:

> A meeting was held on 24 September, 1977, between the Hon. Minister for Petroleum and Power, Dr John P Adiele, and three representatives of an organisation called the Committee of Concerned Citizens, namely Messrs Brown Siekpe, Tari Strongface and Ephraim Fiabara. The meeting, which took place in the Minister's Conference Room, started at 10.30 a.m. Its aim was to discuss certain issues about which the organisation had previously written to the Minister. These were in connection with what the organisation considered 'the survival of the people of the Delta parts of the country'.
>
> It was the view of the organisation that the Federal Military Government had not shown sufficient concern for the conditions of life of the people in that area, despite repeated appeals on the matter by public-spirited Nigerians in the various organs of the media. It demanded, in particular, that two measures be taken by the government to alleviate the suffering of the rural masses in the area. First, that a three-month moratorium be observed on all oil exploration activities until a thorough environmental clean-up was undertaken, at the expense of the oil companies involved, to rid the riverine communities in that area of the extensive pollution which the activities of the companies had caused over vast areas of land and water. Second, that during this moratorium, the flood-gates of the Kwarafa Dam be opened so that large volumes of water could flow down the River Niger for the benefit of the fishing life of the riverine communities.
>
> The discussion between the Minister and the representatives of the CCC was held in a most cordial spirit. The Minister assured them that the Federal Military Government was just as concerned as they were about the situation in the Delta. He, however, pointed out to them certain implications of their demands that would make it difficult for them to be met.
>
> In the first place, stopping all exploration just to clear the Delta of oil pollution would virtually amount to holding the economy of the nation to ransom. Considering the centrality of oil in our economy, and the favourable price of the commodity on the world market, it would be an unfortunate loss of

valuable income to hold up work for three whole months.

Secondly, the Kwarafa Dam had been constructed in such a way as to balance the needs of communities to whose survival the waters of the River Niger are vital, with the necessity for the hydroelectric power which the Dam provides across the nation. To stop the supply of this power for three months would cripple industrial as well as domestic life in the entire country (including the riverine communities) for an unacceptably long time.

However, the Minister informed the Committee that the government intended to set up a high-powered Task Force to study the problems of the Delta communities and make appropriate recommendations. The constitution and terms of reference of the Task Force would be announced in due course.

The meeting rose at 1.40 p.m.

Tolu Adeoye is normally a pleasant and cooperative fellow – you remember how he helped us some two years ago on the controversy surrounding the merger between the National Oil Corporation and Pacific Prospects? But this time he has kept his lips sealed, and that's led me to suspect there's a hell of a lot more to the matter than meets the eye. I haven't been any luckier with those CCC men either – I mean the ones who met with the Minister. Either they're trying to make themselves scarce for some strange reason, or they're all out of town. But to have been conspicuously out of town *all three at the same time* is something that baffles my imagination. You see, I'd hoped I could speak with the men who featured in the meeting so as possibly to confront the facts myself. I didn't want to rely on what the *New Nigerian Times* carried – being practically *owned* by the government, it couldn't be trusted to tell the truth on any serious matter affecting the government. When I failed in my efforts to reach any of these men – of course, I have no access to the Minister, he's a clown anyway – I fell back on Bickerbug.

By the way, I don't think I've told you that he's moved from Okesuna Street to Obalende? Well, that happened about ten days ago. When I asked him why, he said he had more fresh air in his new place than at Okesuna. But I suspect that, since he's waging a war of sorts against the government, he's trying to keep the Security agents confused about his whereabouts. What I just can't understand is why he has chosen to take residence *that* close to Dodan Barracks, if he seriously means to give "the enemy" the slip!

Anyway, I went up to Bickerbug. The first question I asked him was why he was not among the party that met the Minister.

'Come on, Piriye,' he said, 'be serious. Do I look like someone who likes wasting his time?'

He sighed, and rose up from the bed where he'd been sitting. He first walked to the window overlooking the street, pulled the blinds apart and took a quick look up and down the street, then went over to a stack of books at one corner of the room. They were mostly romances – you know, Denise Robins, Mills and Boon, that kind of stuff – but I don't believe he reads any of them. I think it's all a smoke-screen. He fished out one of these books and, opening it, brought out a piece of paper which he held out to me. I looked at him, then at the paper. It was a complimentary card from Frank Segal, the exploration manager of Freland Oil Corporation, one of the newly franchised oil companies prospecting on off-shore locations around Forcados. On the back of the card was a note to Tari – that's Tari Strongface – dated 16 July, thanking him for his "help" and hoping to see him later for dinner at his (Segal's) house.

'Well,' I raised my head to ask Bickerbug, 'what does this mean?'

'That's a good question,' he said. 'We're supposed to be in the same organisation and he never mentioned it to me, even though I'm the leader of the group.'

He was staring at me through that massive bush of hair on his face, his glaring eyes looking rather threatening.

'You seem to think he's playing foul,' I said, 'but this note doesn't tell me very much.'

He shook his head and smiled, then moved to lean against the window. I took a seat, which he hadn't bothered to offer me.

'I admire your innocence, Piriye,' he said. 'But I've seen enough to make me wonder if the shirt I'm wearing now is my own shirt, or an exact replica of it planted in my room. Tari doesn't know I have this card. That's one mistake they make. They don't seem to know I'm smarter than they are. They may think they've got me all wrapped up and ready to deliver. But they'll never get to the point of actually delivering me, because

I'm always one step ahead of them. You know that meeting with the Minister?'

I nodded.

'I told them to go ahead and speak for us. But I knew all along what their real stakes were. I don't think you know this, but I have my sources right within the enemy's camp. After the Minister talked about setting up the Task Force, at the end of the meeting while Tolu Adeoye was writing the draft of the communiqué, our three so-called representatives were busy trying to persuade the minister to include their names in the list. Not because they hoped to use the opportunity to help our cause, but simply to suit their own private and selfish interests. Because the Task Force will have to travel out now and then to assess the state of affairs in the Delta. That means allowances – overnight allowances, out-of-pocket allowances, you name it, that sort of thing. Girls. Plus many chances for the Tari Strongfaces in the group to establish out there the same sort of corrupt arrangements they have here with Frank Segal. Man,' he said, shaking his head and flopping onto his bed, 'I'm fed up with our people. I'll know a proper Beniotu man when I see one.'

That hurt, and I wasn't going to just sit there and take it.

'What's that supposed to mean?' I snapped.

He didn't say anything. He just sat there staring at me, twisting some strands of his beard round his index finger.

'Look here, Ebika,' I told him straight, 'you may think what you like. I know you have problems with your men, but I've also had my own share of troubles. I lost my job for no other reason than that I am a Beniotu man. Right now I'm doing the best I can to survive, in Lagos for that matter. I can assure you of one thing – I feel as deeply concerned about the plight of our people as you do. Perhaps I should also say that I will be as deeply committed as anyone else may be to whatever steps can be taken to improve the situation in our homeland. But please understand that if you feel so free to be suspicious about anyone's genuineness as a Beniotu man, some of us full-blooded Beniotu men also have the liberty to question your right to lead. us. Well, I'll see you later.'

I rose to leave, but he sprang from the bed and held me back.

'Wait, Piriye, please wait,' he said. 'I'm sorry. I didn't mean it that way. If I doubted your genuineness, I wouldn't be telling you all this. Please sit down. I'm sorry.'

I didn't hesitate to sit down. I'd made my point. What the hell! If anyone was going to cast doubts on my Beniotu ethnicity, it certainly shouldn't be a bloody half-caste. Anyway, I sat down on the chair again. He didn't go back to the bed. He just started pacing about, one hand in the pocket of his dirty blue jeans and the other fingering his beard.

'Look, Piriye,' he said, 'I've got plans. I've got plans. By the way, what time is it?'

He was going to rummage through the bed for a watch when I told him, 'Six-thirty.'

'Oh, God,' he said. 'I must go. I have an appointment. When can you come again?

'Any time. I'm jobless,' I joked. 'When do you want me?'

'What about this time, say on Sunday? That's okay for me. But what about you?'

'I'll come.'

'Okay. See you then. I'm sorry to cut you short like this, but I really must keep my appointment. In fact, I'm already half an hour late. But please don't fail. What I'm going to tell you is extremely important.'

'I'll be here,' I promised, then said goodbye to him and left.

Although Bickerbug is a little crazy – crazier than I am, I must admit – there's something (besides my desire to get a story out of him) that draws me to him, and that's his forthrightness. If he doesn't want to talk to you he'll avoid you determinedly. But if he *does* have to talk to you he doesn't mince his words. And I think he means what he says about his mission to save our people from the bad deal they're getting. Which I think is what makes him different from his friends in the CCC. Take their lifestyles, for a start. Bickerbug is the *classic* revolutionary. Not in the fashionable ideologue mould – he doesn't flog the names of Marx or Lenin or Fanon or any of those people. At least I haven't heard him do so and I haven't seen any of their books in his – well, library, if you can call his collection by that name. But he's the true guerrilla type, with a lot more time for planning action and executing it than for the more comfortable ways of doing things. He doesn't shave, wears nearly the same clothes all the time, gets what he wants as fast as he wants it.

But his CCC friends are different. They wear suits and are

clean-shaven and enjoy hanging around the circles of influence and intrigue. They're not very likeable people. Especially that one called Fiabara – God, I hate that son of a bitch. So bloody stuck up. He once boasted to me at one of these NUJ parties, when I asked him if he didn't find the new Minister for Information and Public Enlightenment a little aloof, that he was one of the small handful of people in Lagos with the Head of State's private phone number, that if he wanted any lead on government policy he would dial the Heal of State direct: as he put it, 'When I want to speak to the landlord, I don't go through the janitor.' And the rest of them in the CCC are not much different.

As I said earlier, I haven't seen any of these fellows, so I can't test the strength of Bickerbug's claims against them, though if I have to choose between his word and theirs I'll obviously choose his. What I'm almost certain about is that the CCC as we knew it no longer exists – at least Bickerbug is no longer effectively part of the charade. It's been obvious for some time now, since he started excusing himself from their moves and they staying away from his. Any agitation or rally you see advertised today is almost certain to be a solo Bickerbug show, though the Committee will have initiated the publicity. My only worry about the growing separation is what Bickerbug might do. I tell you, I think that boy is planning something big, something crazy – something he's probably certain the rest of the Committee won't join him in. He hasn't told me yet, but I'm hoping it'll be what he wants to discuss with me when we meet on Sunday.

I don't think Priboye is going back your way for some time. He tells me he's expecting certain shipments and would like to take delivery of them himself at the ports before returning to Warri. I'll simply keep on writing these notes as the need arises and pile them up until he's ready to travel. So don't be surprised if he comes home with a stack. If you have abjured newspapers, you may at least tolerate a journalist's situation reports!

Cheerfully,
Piriye

Lagos, 18.10.77

Dear Tonwe,

Bickerbug didn't keep our date. I went to his place at 6.30 p.m. two days ago as agreed, but – to use a popular phrase – I met his absence. The door wasn't even locked, and the room looked a little ruffled, though given the man's lifestyle I wasn't so terribly surprised. I've asked around but no one seems to know where he is. Or wants to tell me.

But those three jokers in the CCC are back in town – assuming they ever left it. I got the first hint from Lati (that's Miss Ogedengbe of the *Chronicle*, you remember?). I called her at the office to find out if anything was happening at their end, and she greeted me with the news.

'Your friends were here today,' she said.

'Who? What friends?' I asked.

'Of the three Cs', she whispered, barely audibly.

'Is that right? What happened?'

'Mr P,' – that's what she calls me – 'are you free this afternoon?'

'I'm jobless!'

'Oh, I don't mean that! But, um – where are you going to be about 2.30 p.m.? I've got news,' she said, teasingly.

'I could buy you a snack on Broad Street,' I offered.

'Okay. See you at the junction of Broad Street and Martins.'

Roughly two hours later I took a taxi from Adeniran Ogunsanya and headed for the Island. (By the way, my car's grounded and I can't afford to fix it.) As I walked from Tinubu Square down Broad Street who did I see just outside the Cathay Chinese but Brown Siekpe of the CCC, in his usual suit and briefcase, pacing rather hurriedly.

'Hey, Brown,' I greeted, pulling his arm.

'Oh, hello Piriye. How are things, man?' he said, without really stopping.

'All right.' In view of his urgency, I didn't seem to have very much room for more courtesies. 'Where's Bickerbug these days?'

'I don't know. I've been looking for him myself. If you don't mind, I'm in a bit of a hurry. I'll see you some other time.' He had already turned the other way.

'No problem,' I waved him off. Bastard! I don't like people

who act so important. He may have been going to the Central Bank for tea with the Governor, so what?

I turned and moved on to keep my date with Lati. I was a little late when I got to the intersection of Broad Street and Martins Street. Lati was waiting by the Lufthansa office, and when she saw me she greeted me with that warm toothy smile that has become something of a trade mark of hers. Interesting girl – whether she's bringing you cheering news or bad news she starts off with a smile! Anyway, I greeted her with a friendly pat on the cheek and we walked up Martins, past the airline offices and up to the Phoenicia Restaurant. I once had a fantastic chicken sandwich there and thought I would treat Lati to one. She said that would be great, she was starving. When the steward came I ordered two sandwiches and two Cokes.

'So how's the old place?' I asked as we were waiting for the orders.

'Not bad,' and she smiled again. 'It's been a dull day though. I was writing my report on the Senegalese troupe participating in FESTAC '77. It wasn't going very well, so when you telephoned it was such a relief to set aside the assignment and answer your call. But I'll finish writing the story today before I leave the office.'

She looked around, admiring the restaurant. I looked her straight in the eye and hoped she would appreciate the urgency of my request without my actually making it. She just smiled and looked away. Women – playing games with everything!

'Well,' I said, 'what are those clowns up to? What are they doing at the *Chronicle*?'

'Well,' she sighed, 'I'll show you.'

She put her hand in her handbag and started fishing around for something.

'By the way,' I said, 'I just ran into Brown Siekpe on Broad Street. He was in a real hurry.'

She nodded, her hand still in the bag.

'He was wearing an ash-coloured suit with a red tie,' she recollected, 'and carrying a brown Samsonite briefcase?'

'Yes,' I nodded. 'He hardly stopped to return my compliments. What's he up to?'

Lati smiled and shook her head. She brought out a piece of paper from her bag and was about to hand it to me. Just then the steward came up with a tray bearing the sandwiches and Cokes, and she dropped the paper back in her bag. I was burning with curiosity while the steward cleaned our table with a cloth and placed the snacks in front of us. I barely noticed when he slipped the bill beside my plate and walked away. I was staring at Lati's bag with the anxiety of a little boy expecting a Christmas gift.

'Take a look at this,' she said as she handed me the paper. 'It was brought in earlier today by Siekpe and, er ... the very tall one – what's his name again?'

'Ephraim Fiabara.'

'That's right. I was with the editor when they both came in. It's a statement they wanted published in tomorrow's issue.' She bit into her sandwich.

I read feverishly through the typed statement. It was jointly signed by the three men, and in it they said they were aware of the efforts of certain individuals – unnamed, of course – to use the name of the Committee as an umbrella for perpetrating acts of violence, ostensibly in their bid to save the people of the Delta from the problems which an insensitive government had brought on them. The Committee wished to assure the Federal Military Government and the general public that they were committed to seeking a peaceful solution to the Delta problem. They were peace-loving citizens and would like to dissociate themselves totally from any acts of violence planned or perpetrated by any person or persons whatsoever, whether singly or jointly.

I was a little worried after I read that piece, because I wondered if it had anything to do with my not seeing Bickerbug. Believe me I'm still worried, because it may explain why everyone is trying to hide the facts of his whereabouts from me, and Siekpe's behaviour that afternoon has done little to soothe my fears about the poor fellow.

'So this is coming out tomorrow?' I asked, when I had finished contemplating Bickerbug's whereabouts.

'Not any more,' Lati said. 'They didn't stay very long with the editor. I had left his room and returned to my desk. But I kept an eye on their movements. Not more than ten minutes after they came into our offices they went out again. The editor passed the copy on to Maduwesi to prepare the story for production. But shortly after I got your telephone call about noon, one of the three men – I don't know which – called in to tell the editor to withdraw the material from production.'

'Why?'

'He didn't really explain. Something about orders from higher quarters. He said he'd get in touch later.'

I found it all quite curious. The postures of those men in the CCC were nothing new to me, but the business about "higher quarters" was introducing a certain complication to the problem, and this was what got me really worried about poor Bickerbug.

'How did you come across all this?' I asked Lati.

'What – the paper, or the information?'

'Both,' I said.

'Well, I have my ways,' she intoned coyly, and laughed. I couldn't help smiling myself. 'Actually, Maduwesi was quite cooperative. But he doesn't know that I stole the statement and made a xerox copy. Do you want to keep it?'

'Oh, that's very kind of you.'

'Not really. It'll cost ten kobo,' she said, stretching out her hand. Then she laughed again and said, 'Just fooling. Come on, finish up your lunch. I've got to get back and complete that FESTAC story.'

I had barely touched my sandwich, while she was almost at the bottom of her Coke. The atmosphere was getting a little too serious and I tried to relax it.

'Well, how's the old place?'

'How many times do I have to tell you?' She pretended to be exasperated. '*Not bad!* Relax, Mr P. I think you're allowing this thing to get to you.'

'I know.'

I finished the sandwich and started on the Coke.

'Well, what have you been doing with your time?'

'Same as you. I just did a FESTAC story.'

She smiled at the coincidence.

'Will you give it to us?' she joked.

'You know better than that,' I said. 'No way. It's for the *Daily Republic*. I won't have a damn thing to do with the *Chronicle* any more.'

'Including its reporters?'

She had a mischievous smile at the corner of her mouth.

'Oh no, Lati. You're okay,' I said, looking straight into her eyes. She smiled and looked away. There was some silence while I finished my Coke. Shortly afterwards, I beckoned to the steward to take the money and keep the change.

The statement from the CCC will probably never get published. My suspicion is that the withdrawal was requested by the National Security Service, though how they got to know about it, and to what extent they are privy to the moves of the CCC, I can't say. But even if the statement doesn't get published, it will provide part of the valuable evidence we will need as we construct what I consider an important segment of the book – a chapter or more perhaps, on the penetration by government and its agents, in an undercover manner, into the activities of various persons and organisations having to do with the situation in the Delta. I know how you feel about the conception of titles at this stage of the project, but a title like 'The Eagle's Shadow' wouldn't be too far-fetched for such a chapter.

Anyway, Lati must have grown somewhat perturbed by my silence as we walked away from the Phoenicia in the direction of the Kingsway Stores. I vaguely recall her thanking me for the lunch, and I think I must have said something like 'Don't mention it.' What I do recall vividly was her asking me, 'How's your wife?' Our eyes met at that point and all I said in reply was 'Okay,' because I didn't want to discuss an issue that has become more or less common knowledge among those with whom I've worked. By this time we had got to Kingsway and while I headed for the bus stop she took her leave to return to the *Chronicle's* premises just around the corner. Terrific girl. A healthy curiosity, delightful sense of humour, etc., etc.

I must return to the issue of my wife not simply because the matter crops up here but especially because in your last letter you devoted a certain amount of space to urging a reconciliation between us. I know you are a peace-loving man, but I think you are a little out of tune with the situation I am dealing with here. How would you feel if your wife didn't give a damn about you, wouldn't cook you a meal let alone put any of your clothes in the wash, felt too big to even wish you a good morning, thought you had no right to question anything she said or did, and several other things I won't even bother to mention? When my friends come to visit me she simply walks across the living room without even giving them a simple hello. Is that the sort of woman you would want me to be reconciled with, when she has proved herself a much worse tragedy to me than the retrenchment I suffered from the *Chronicle*? If we had had children – you're damn right, that's the real problem – I could have concentrated on them whatever love and attention is within me to give, and I would have been content to ignore her presence here quite blissfully. But I see her here every day, and she has become like a bloody thorn in my side. Frankly, I wish she would pack her goddamned things and leave. Instead, you urge that I should allow our Beniotu elders and relatives here in Lagos to come over and settle the issue between us.

Now I know you mean well and that, despite your guarded feelings about what you consider to be my aggressive Beniotu nationalism, you would be happy to see me maintain my union with my Beniotu wife in the good old traditional way. But why must I pay such a heavy price for this? Has she given me a child that will help keep alive the Beniotu thing between us? You know, she had the nerve to brag to me once that she had to have an abortion before she met me. So the fault is mine that we are childless! Before we got married, people never tired of telling me that she came from a good Beniotu family. I know the Koripamos are a good family, but what good in hell has that been? Has it stopped their child from being a complete bitch to her husband? People talk about a family from which a girl comes, as though the family created her innate nature or as though the family is going to direct her conduct with the man she comes to marry. Frankly, I've seen enough of life with Tonye not to believe any of that bullshit. And I am perfectly convinced I can carry on my Beniotu nationalism without having to be saddled with a bloody Beniotu vixen. She has messed up my life, soured my outlook, my life-style, my everything! If those elders and relatives you invoke really want to do us any good, they should come and take her away from here.

But enough of that. I intend to go looking for Bickerbug after this. If I find out anything, I'll let you know.

Greetings,
Piriye.

2
BILLOW

Oh, harken to my voice, Alowei my love
For I can cry no more
Here I stand pouring tears by the
    waterside at Odoni
The waters have swelled to my lips
I need no cup to drink from now
For mine overflows with my tears
So harken to my voice, Alowei my love
For I can cry no more.

Ijo folk song

Lagos, 30.10.77

Dear Tonwe,

When I called on Priboye two days ago, I was told he had gone to Warri. So I couldn't give him the letters I'd previously written. But I must write this one anyway, and they can all go together the next time Priboye goes home from here.

I found Bickerbug – or more correctly, I know where he is or can be reached. But, I tell you, the last few days have been rough and have taught me a sobering lesson or two. My wife has also gone, thank God!

Here's what happened. After I'd written my last letter to you, I took a taxi to Obalende to look up Bickerbug once again. I knocked and knocked, but there was no response. I turned the door knob, and the door flew open. The room was in the same state as last I found it – bed ruffled up, as of an occupant who'd just woken from sleep, floor dusty and littered with books and papers, etc. I sat on the bed for a while, and to while away the time I proceeded to leaf through some of the books, since there was little else in the room anyway. As I told you in a previous letter, there wasn't much there – a few books of popular fiction, one or two volumes of poetry, one or two novels by Chinua Achebe, Alan Paton's *Cry the Beloved Country*, a book of plays by J.P. Clark, and so on. All these weren't surprising – the fellow read English at Nsukka, as far as I can remember. There were also a few other books – Alagoa's *History of the Niger Delta*, a Beniotu primer, some Nigerian geography books, and one or two magazines. The only strangers in the batch were a book on civil engineering and another on chemistry. I couldn't think how they got there – but the fellow is crazy and seems to pick up junk from all kinds of places.

After a while I got tired of sitting there and waiting for no one, so I shut the door and left. I hadn't gone more than a few paces when someone hissed to me. I looked round, and there was a woman standing by the corner of a nearby house and beckoning to me – middle-aged, matronly, and somewhat timorous. I walked towards her, wondering what the hell she wanted with me. The diminishing light of the evening put her in a rather mysterious light.

'Good evening, madam,' I greeted, bemused.

She didn't return my greeting. Instead, she beckoned me on into her house. As we got in, she shut the door quietly and asked me to sit down. The well-upholstered furniture and the deep rug belied the exterior of this colonial-style bungalow. As I took my seat I found myself facing an elderly man, who turned out to be her husband. A somewhat enlightened pensioner, he seemed to me.

'Good evening, sir,' I greeted again.

'You are welcome, my friend,' he returned in low deep tones. 'I am sorry we have taken you by surprise like this. We will not ask you your name and we will not tell you ours. We have seen you come to visit your bearded friend, our neighbour, a couple of times and we thought we could help. But we really don't want to get involved in anything, you understand?'

I nodded. I was getting a little worried. I looked at the old man, then at his wife.

'What happened to him?' I asked the man.

'About a week ago,' he said, 'two men came to his place. They had parked their car on the other side of the street. When they walked past our yard to his door we didn't think anything of their visit – we simply thought they were the usual sort of visitors that called on him. They wore plain clothes. Not long after, however, there was some shouting in there, and then some scuffling. There must have been a heated argument between your friend and his guests, and I think there may have been some violence. After a while the place became quiet again. And then the door opened, and the two men came out with your friend. Are you listening?' he asked, apparently because I had been looking at the floor and seemed to be lost in thought.

'Yes, I am,' I quickly replied, raising my eyes once again to him.

'As they walked him to their car, they seemed to be restraining him, and we heard him saying something like, "You'll never get me to admit anything. I demand to speak to a lawyer," and they in turn said something like, "Don't worry. You'll be given one. Just come along quietly and you won't get hurt." They drove him away in their car, and he hasn't been back since then.'

I nodded a few times, and reflected for a while. The bastards, I thought. What have they got on him? Who's been talking to them?

'What time of day did this happen?' I asked the old man.

'Er ...'

'About four or four thirty p.m.,' his wife chipped in. 'I had just come home from my shop.'

'I see.'

'They didn't bother to close the door when they left,' she said. 'I was the one who went across to shut it when night fell.'

'I see,' I said. 'Do you have any idea where they could have taken him?'

'Not really,' said the man. 'We suspect they were plainclothes policemen. But where they took him to, we can't say.'

I rose slowly to leave.

'I'm sorry, we never offered you anything,' he said. 'Would you like a beer or something?'

'No, nothing, thanks, sir. Maybe some other time. I really must go and look for him.'

As I walked towards the door he rose from his chair and walked after me.

'You know,' he said, 'I've never really known your friend. I've seen his picture once or twice in the papers, especially after that time they arrested him at Campos Square. And I spoke to him once after he got home from school one day. He seemed to me a very intelligent and very energetic young man. I admire his thinking about the situation in the country today – I had a bit of the radical spirit in my youth, you know, so I understand how he feels. But please try to talk to him if you can. The times we are living in are not exactly the best of times. The atmosphere is not really conducive to his style of approach – this is not a civilian regime, you know, and even civilians can be just as dangerous. So please, try to get him to cool down. Talk to him, okay?'

'I will,' I said. What else could I say?

'Does he have any family?' asked the wife.

'You mean, like a wife and children?' I said.

She nodded.

'Not that I know of.'

'*O ma se o*,' she said sadly.

I wanted to ask her what the hell that had to do with anything, but I decided not to waste my time on a useless argument. As I turned to step out of the house the old man grabbed me by the arm.

'Remember,' he whispered, 'you didn't hear any of this from us. You don't know us, and we don't know you. Okay?'

'Thanks for your help,' I said, snatching my arm from his paw. 'Goodnight.'

Rather than say anything in reply they simply waved at me. I opened the door, and as I stepped out I shut it with a lot more publicity than the woman had done earlier. Not that I wanted to offend them or show any ingratitude as such, although I'm not sure I react very nicely to people who are eager to guard their comforts while others are being brutalised. Besides, the woman's feelings about having a wife and children were still fresh in my mind. But I didn't *deliberately* slam the door. I suppose I was just disturbed or confused or something. And I wasn't quite myself as I walked out across the street.

Well, I'm not so stupid as not to know you have to be careful how you deal with Security. I'd of course been to Police Headquarters quite a few times before – you know only too well how most reporters at the *Chronicle* hated to be assigned to the "McEwen Square beat", not because they did anything to you there, but because the police are experts at not really telling you anything. But the Security section is in a class by itself. I've now learnt that you don't just go there wanting to find out if anyone is being held. Who told you he's being held there? What's your connection with him? How much do you know of the matter for which he's being held? That sort of thing. I may have suspected all this. But I was prepared to take the chance. I thought it might well be that my identity as a journalist would protect me from the sort of risks that an ordinary citizen would face, or at least it might temper the wrath of the authorities over my "impudence" – after all, they wouldn't really want to be embarrassed by an outcry from the Union of Journalists against the arrest of one of its members!

Anyhow, to be on the safe side I thought I should tell someone what I was about to do. When I left Bickerbug's neighbours I took a taxi straight to Lati's place on Herbert Macaulay Street, Yaba – she lives with her aunt, Mrs Aina Kumolu-Davies, a quite influential Lagos socialite and a rich businesswoman. Yes, I thought, Lati should know.

'Shall I come with you?' she asked eagerly, after I'd told her my plans.

'What on earth for?' I asked.

'Please, Mr P. I know Bickerbug is your beat, but please don't deny me the excitement. I won't steal your story. I just want to see what's happening.'

'Excitement!' I said. 'I admire your guts, girl, but this is not quite your usual kind of excitement. It's dangerous. Anything could happen to me. I've come to tell you so that, if I got into any trouble, there would be someone who'd know where I was going.'

She could see I wasn't fooling.

'Okay,' she said, subdued.

'And Bickerbug isn't just a "beat" to me,' I continued, after a brief silence. 'He's Beniotu, you know, and so am I. He happens to be fighting a cause which is very close to my heart, the welfare of our people. Calling him a "beat" implies that I simply want to get a newspaper story out of his troubles. Well, he means a lot more to me than that. After all I've been through at the *Chronicle* I'd be a fool to treat my people's welfare as just an item for a newspaper report.'

That must have got through to her, because at that point she sighed and sank down on a chair.

'Oh, I'm sorry, Mr P,' she said in a low, tremulous voice. 'You know I didn't mean any harm. I just ... I just .... I don't know. I think it just came out of me. Or maybe I just wanted to go with you. I know I'm not from your place, you don't have to tell me that. But I know what's going on there and I think it's awful. If I didn't care about it and about how you feel, do you think I would cooperate with you as much as I've done?'

'Well, I appreciate what you've done...'

'I'm not claiming any credit or demanding gratitude. I just want you to know I care too. I don't have to come from your place to see the terrible injustice being done. Anyhow,' and she rose cheerily, 'you go ahead and see Bickerbug. I'll keep a lookout for you, okay?'

I hate to admit it, but she had me at a loss for words!

Perhaps I should have said that the first thing that drove me to tell Lati about my visit to Security was the tip she'd given me about the call at the *Chronicle* by those CCC men – Brown Siekpe and Ephraim Fiabara. It had seemed to me there was a connection between their withdrawal of that statement from the *Chronicle* and Bickerbug's disappearance. Those individuals they claimed in their report were trying to commit acts of violence under the cover of the CCC – could they be referring to Bickerbug? Was there something Bickerbug had got involved in that he didn't want to tell me about? And what about those plans he said he had and could have told me had we met on the 16th as planned? In any case, I thought it made sense to go back to Lati as the source of the information which might help me make sense of the background to Bickerbug's arrest by the NSS agents.

I hope, after all I told you in my last letter of the deterioration of relations between my wife and me, that you will not be so unfair to me as to wonder why, if I desperately wanted someone to know where I was going to be, I didn't think of her first. Perhaps by the time you get through reading this letter you will understand why the question should never have arisen in your mind.

I got to the Police Headquarters at 10 a.m. the next day. I'd dressed respectably enough – a light blue shirt, navy blue trousers, a pin-striped dark grey Tootal tie, an ashcoloured jacket whose sleeves rested just above my gold-plated cufflinks, a pair of decent black shoes. I'd felt that in this outfit I would gain more respect from the police than if I was to appear as a scruffy-looking pen-and-paper reporter.

When I got into Security, I was surprised to see that it looked like any other police establishment. There was a long counter which led into a broad hall where several police officers – male and female – were seated at desks. Beside the counter, two police officers were having a loud cheery talk with a heavy fellow in *agbada* – I think he was a businessman, from what they were saying, and I heard them call him "Alhaji" a couple of times. The policeman at the counter was dozing, with his head and left arm rested on the dark formica top of the counter, and his mouth open. The two officers chatting with the Alhaji didn't seem to take any notice of me, so absorbed they were in their conversation, although I thought I commanded enough presence to be noticed. I adjusted my tie. I was about to knock on the counter to rouse the sleeping policeman when a food vendor with a basin of food on her head lifted the barrier at the entrance and called a greeting as she went in.

'Officer.'

One of the policemen chatting with Alhaji turned to her and

replied, 'Ah, Mama Iyabo. How now?'
'E go good.'
'Make you leave my own for table,' he instructed.
'Okay. Corp'l,' she called to the sleeping policeman, as she replaced the barrier with a loud crash. 'Morni' oh.'
He roused himself, opened his reddish eyes, and stretched his arms.
'Aha. Iyawo, you done come?' he replied. He didn't seem to have noticed me yet. He looked under the counter and handed her a lidded plate with a spoon, saying 'One naira. No meat.'
While the food vendor put down her basin and was dishing out the order for rice, the policeman finally turned to me and said:
'Aha. Oga, can I help you?'
'Yes,' I said. 'Um, I believe that one of my friends is being questioned by the police here. I don't know exactly which officers are in charge of the case. He hasn't been home for a few days, and I thought perhaps I should come and find out if he's being held here.'
He took his plate back from the food woman, and paid her. She put her basin on her head and walked on.
'You are Mr ...?'
'Dukumo is my name. Piriye Dukumo.'
He stared at me briefly, obviously trying to figure out the next question. Or maybe I looked familiar to him?
'Of er ...?'
'Pardon?' I leaned slightly towards him.
'I mean, where are you from? Where do you work?'
'Oh,' I straightened up. 'I'm from the press.'
'M-hm,' he grunted, running his reddish eyes down my figure. 'What is your friend's name?
'Mr Ebika Harrison.'
'I see,' he said. 'Wait here. I'll go and find out.'
He had a naturally dull look on his face. But at this point there was an added glower to it as he rose from his seat and went over to the office behind him. I couldn't decide whether he was reacting to the mention of Bickerbug's name or to my holding up his breakfast. I didn't really give a damn which. I turned my back to the counter and leaned on it to look at the more pleasant view outside. There were flies dancing all over the counter, and cobwebs in every corner of the place.

The two other policemen had seen off the Alhaji and now passed back in without even saying a word to me. The traffic of callers here was lean – as I said earlier, the NSS section isn't a place that you just get up and go to. It's not like the regular charge office, where you find people coming in and out with complaints and all forms of civil cases – taxi drivers prostrating themselves on the floor and begging passionately to have their papers given back to them; or brawlers with cuts and bruises on their faces, hauled in from drinking bars the night before; or victims of armed robbery who have lost their most valued possessions and barely escaped with their lives. It struck me that, probably due to the absence of these routine forms of daily encounter, the policemen here were somewhat less sour faced than the ones you meet at the charge offices. As I stood waiting for the corporal, other policemen came in and went out and, although they were not particularly polite, they didn't look at me as though I was a condemned or habitual criminal.

My mind turned to Bickerbug, and I started wondering what could be happening to him and where they had put him. From what his neighbour had told me, he must have had a nasty brush with the plainclothesmen who came to take him away. Although the policemen I saw here didn't look particularly sour, policemen are still policemen and don't like anyone like Bickerbug who refuses to "cooperate" with them, as they like to put it. And Bickerbug is a particularly tough nut. What was it they were trying to get him to admit? I knew how much seeing him meant to me, but was it wise for me to get involved in whatever mess he might be in? Well, what the hell, I thought. If I backed out now I would be guilty of cowardice – all the corporal had told me to do was to wait – and, what's more, a rather ignoble betrayal of a man fighting my people's just cause.

At this point the corporal returned to the counter. I turned round to face him, but he didn't say anything. I was going to ask him what information he had for me, when I saw another officer approaching me.

'Yes, my friend,' he said, 'can I help you?'

His shoulders were heavy and menacing, and his thick neck seemed to sink between them.

'Good morning,' I replied. 'I'm trying to find out if Mr Ebika Harrison is being held here for questioning. I've asked all over for him and nobody knows where he is.'

'So you think he is here?'

'Well, I'm not sure. But he never told me he was travelling. And I know he's had trouble with the police before.'

He ran his eyes over me.

'You know him very well, do you?'

I hesitated a bit, then answered, 'Yes, I know him.'

'He's your friend?'

'Well,' I hesitated again, 'everbody knows him.'

'But you're the only one who has come looking for him here.'

I didn't know what to say to that. But I didn't want to appear overcome either – that's the last thing one should do with the police, as you know. I simply had to take care not to put my foot in my mouth. He surveyed me once again.

'What's your name, sir?'

I told him.

'Come with me, please,' he said, turning to go back inside.

For a while I stood there. It was like being invited by a lion into his den. I wasn't quite prepared for this. I wanted to ask him to let me explain a little further, but he had turned the bend towards the line of offices from which he had come. I had no choice but to lift the barrier and amble my way in, so confused I nearly knocked down the rice vending woman as she made her way out.

I was ushered into an empty room. It was large and had no curtains on either of its two windows, only louvres which were closed. There was one plastic chair in the centre of the room, and there the officer made me sit. As I sat there I was beginning to feel like a criminal. It seemed to me I had lost all dignity the moment I crossed that barrier – and now my safety too. This was obviously an interrogation room. The officer who had brought me here had disappeared, leaving me staring across the large empty space in which I had been trapped. There wasn't even a telephone there. On the wall above the closed door there was a grey, rectangular-shaped metal box from which a wire led out through the top right edge of the door. The box had slits, and was connected to a red ball-like object. I felt truly cornered, but I was determined to keep my cool.

So many thoughts flashed through my mind as I sat on that chair – far too many and too confused for me to reconstruct. I believe basically I was trying to prepare myself by imagining all the worst possible things they could do to me – not excluding torture and imprisonment without trial – though I couldn't bring myself to imagine what in damnation I'd done to merit such a savage fate. I must have been sitting there for over an hour unattended – I couldn't understand what the hell was going on. Once in a while someone would stare through the window overlooking the corridor – or two people would peep in together and go away talking. I kept wondering what on earth I was being held for. Were they searching their records to see if they had anything on me? Or could they be going through my newspaper articles – assuming they were diligent enough to keep copies of them – to see any possible connections between me and whatever it was they had on Bickerbug? Or could it be ...? I just couldn't make sense of the whole game. After a while I got rather hot – whether from so much confused thinking or because the door and windows were shut, I'm not sure now. There was no ceiling fan or anything like that. Since nobody was there to help out, I got up from the chair and opened the windows at the back, overlooking the premises. For a while I stood there, taking advantage of the little air that came through, for there was hardly any draft.

After some time I heard the door open, and I returned to my seat. It wasn't the officer who had brought me in. This time it was an older one – fiftyish, I would say. He surprised me with the smile he wore on his face. He was clearly a Northerner – very dark, three marks on his forehead, perhaps from the northeast or thereabouts, but with an unmistakeable Northern accent anyway.

'Good day, Mr Dukumo,' he beamed at me, showing his teeth.

'Good day, sir.'

We shook hands. He motioned me back to my seat. A junior officer brought him a chair, and he took his seat a few paces from me.

'So, how is journalism these days?' he asked, still beaming.

I said it was okay. I really wasn't in the mood for pleasantries in this place where I'd been held long enough against my will.

'I like your articles, you know. Especially the one you did recently about er ... Remind me, what was it again?'

'The one on power cuts?' I offered.

'No, not that one. Er ...'

'Labour unions?'

'No. Something about the Delta, I think.'

'Oh, the one on gas flaring?'

'That's the one,' he said eagerly. 'I *loved* that piece. It was very beautifully written. And you know, I agree entirely with you. How can we have so much wealth in the country and be burning it away? It doesn't make sense. You were *so* right.'

I nodded, but didn't say a word. I had begun to distrust this solidarity. Was this why I'd been kept here for over an hour in this bloody awful unventilated room?

'So, I hear you have come to pay us a visit.'

'Not at all, sir,' I corrected him quickly. 'I only came to find out if Mr Ebika Harrison is being held here. I didn't ...'

'Ah, of course. You know him well.'

'Well ...'

'No, that's all right. Nothing wrong in knowing him. After all, he's a man and should have friends,' he chuckled. 'And why shouldn't a man's friends look out for him if they think he's in trouble, or haven't seen him for some time?'

I didn't say anything.

'Tell me, Mr Dukumo, what do you think of Mr Harrison, or Bickerbug, as he is popularly known? What do you think of him? That's an unfair question to ask a man of his friend. But, this is just between us. What do you think of him and his approach to things?'

'Let me get something clear, sir,' I said. 'Is he actually being held here?'

'Er,' he thought for a while, 'y–yes, I should *think* so. Of course, so many things happen here every day and I can't possibly know them all. But let's assume he's here. What do you think of his methods?'

'Like which, sir?'

'Come on, Mr Dukumo, you know what I mean. The man has obviously declared war on the government. He knows there is a ban on public speeches, especially those of an inflammatory nature, and he still goes on making them. Denouncing the government. Calling everybody names. But, you know, the most disturbing aspect of it all is the ... the ... the *ethnic* character of his agitations. Do you think it's right?'

'How do you mean?'

'I mean, it's always the *Beniotu* this or the *Beniotu* that, the

Beniotu people this, the *Beniotu* nation, the *Beniotu* demand this and that. I mean, why – why all that kind of talk?'

I didn't say anything. I didn't want to. I could guess which way he was going, and I didn't want him to lead me in that direction. I simply sighed and shook my head, pretending to be just as baffled as he himself was pretending to be.

'Mr Dukumo,' he said, breaking the silence, 'you are Beniotu yourself, I suppose?'

'Yes, I am,' I said, looking him in the eye.

'And, er, do you support this sectionalism?'

'Well, sir, it's hard to say. This country is made up of so many ethnic groups, each trying to protect its interests. Sometimes the approach is a bit harsh, but some time or other we all have to identify with our people and their aspirations.'

He chuckled, flashing his teeth again and fixing his eyes on me.

'That's a clever one,' he said. 'You writers have a way with words. You know, I remember in that article you wrote about gas flaring, you said something about the waste and the – I think you used words like "inhuman exploitation" of the owners of the territory containing the oil, and you spoke with special emphasis on the Beniotu people. Do I take it you feel very much the same way, perhaps, as Bickerbug does?'

I knew he was leading up to that.

'Well,' I said, 'everybody has his own style.'

'But you all have the same objectives, is it not so?' By this time his face was beginning to tighten a little. 'Tell me, Mr Dukumo,' he went on, 'do you think any section of this country, however aggrieved it may be about the way it thinks it is being treated, has any justification in adopting violence as a means of protest?'

'Officer,' I said, 'I really have no idea what you are talking about.'

Frankly, I was tired, hot, tense, impatient. He stared me in the eye for what must have been a good thirty seconds. By now he had lost all the smile on his face.

'You don't?' he asked.

'No, I don't,' I replied.

He stared me once more, sighed, and rose from his seat.

'All right,' he said. 'I'll give you time to think about it. I'll be back with you soon.'

He left the room, and he never returned. Again I was left to wait. I looked at my watch, and it was about 1 p.m. This time I sat there for more or less three hours, because by the time anyone came into the room again it was way past 3.30 p.m., and many of the staff had passed by my window on their way home. As I said, the officer – I never got to know any of their names – didn't come back to the room. At nearly 4 p.m. another officer, junior to my interrogator and from his accent a Diala man, came into the room and said to me:

'Oga, let's go.'

I rose in disbelief. I was perplexed. He didn't wait; he just moved on like one expecting a person to follow without question. I sprang quickly after.

'Where to?' I asked.

'You don't want to go home?'

My face and my heart relaxed briefly, but only until I saw him turn into one of the offices. I wasn't sure whether or not to follow him in. If he really meant I should go home, then it was more appropriate that I should continue along the corridor until I went out through the barrier at the counter. Wouldn't you do the same if you were in my shoes? I had proceeded to do just that when I was stopped.

'Hey, oga!' the officer shouted

I walked back to him and said, 'Yes?'

'Come and sign this form,' he said, pushing a long printed sheet of paper and a biro before me on a table.

I settled down on a chair near the table, and ran my eyes nervously over the paper. It seemed a routine enough form, with spaces for you to put down your name, home and office addresses, occupation, next of kin, and so on and so forth. But I had a basic objection to signing that form. I had not been brought to Police Headquarters under arrest or even by invitation from the police. I had come there of my own accord, and though I had been subjected to a most irritating period of waiting and interrogation for nearly the entire working day, I had done nothing for which I should be written into the police record books.

'Why must I sign this?' I asked the man.

He was bringing out his things from a cupboard against the wall and getting ready to go home. When he heard my question he stopped what he was doing and looked at me in disbelief, or disdain – I wasn't sure which.

'What did you say?'

'I said, why must I sign this?'

'Why don't you ask the DSP?' he said. 'He asked me to give it to you. I'm only doing my job.' He returned to his packing.

I held the paper in my hand for some time. It was clear that any further questions from me, not to say any attempt to resist filling the form, would be a total waste of time. I was right inside their premises and they might think nothing of keeping me overnight there – perhaps moving me to the quarters for unfriendly guests – if I indicated I needed more time to consider whether or not to sign that document. I was still contemplating the paper and weighing up all the possible implications when the ASP came over to my side.

'Look, my friend,' he said, 'I have to go to the mechanic and see about my car. So please make up your mind.'

Well, I bent over and filled out the sheet. He took it from me, looked it over to be sure I'd done what I was supposed to do, and said, 'Thank you. You are free to go.'

Tonwe, I don't know how close you've been to the police. I recall you were summoned as state witness once in the hearings over the Pacific Prospects controversy. But have you ever been inside their premises as a guest undergoing the various forms of routine they normally put their guests through? If you have, then you'll understand how I felt at the point I was declared free to leave.

That reminds me. I'll tell you one major reason I had my hesitations over signing that paper. Someone told me once – I can't remember who now nor how he got to know – but someone once told me that the last thing a man should ever hope for was for his name to get on the Security list, because it would never leave it. From one regime to the next, whether civilian or military, your name would continue to be there – the list is never erased or revised, only updated. They may not touch you or bother you, but you can be sure you are being closely watched. The only way your name leaves the list is if a new regime gives you an appointment as an ambassador or a commissioner, or something at the level that brings you close enough to the controllers of power to enable you to influence the removal of your name from the list. So you can see why I would have given anything to avoid being written into their records and why I felt only a partial relief when that Diala officer finally said I could go home.

I say partial, because the knowledge that I was now bound to them by that blasted document left me with a certain sense of unfulfilment, and reminded me of the job I was there to do in the first place. My mind returned to Bickerbug and to our project. Surely he must be somewhere around here. I thought I could make one last desperate (ill-advised?) effort to locate him and if possible talk to him. I had gone a few paces out of the office when I returned to ask the officer, who by this time had grabbed his things and donned his cap and was on his way out.

'Er, officer, is there any chance I can at least see Mr Harrison, even for one second?'

'Why didn't you ask the DSP?' he replied. He didn't stop to say anything else. He just stormed past me into the street.

I got out and took a taxi. At first I told the driver to take me to Herbert Macaulay, so that at least Lati could see I was back in one piece. But on second thoughts I decided to go home first. Why? The DSP seemed to have put a certain emphasis on that article I had published in the *Daily Standard* about gas flaring – at least the interrogation had ended rather abruptly on that subject. It's funny how things we say in our inspired moments come back at us with all too stark consequences. Of course I was driven to a large degree by my nationalistic feelings as I wrote that piece, but I didn't think it was sufficiently inflammatory nor did I suspect for one moment that the police would remember it. I never gave them credit for such efficiency. Anyhow, I became anxious to look at that article again, so I asked the driver to take me home.

I got home at about 5 p.m. My wife was not there. I wasn't at all surprised, although she usually got home from work about 4. But on walking through the living-room towards the bedroom I saw a note on the dining-table. I picked it up, and it read, 'Goodbye, Piriye. I'll see you in court.'

My eyes quickly surveyed the living-room. She had taken away her flower vases, the hand-poofs and head-rests on the cushion chairs, even some of our wedding photographs on the wall, and pelmets. I moved on into her room – we'd been sleeping in two separate bedrooms for several years now – and it was practically empty but for the basic furniture. So I knew Tonye had gone for good. I didn't bother to look elsewhere, like the bathroom and the kitchen, for further evidence of her leaving. I knew she had gone.

As I sat down in the living-room, I wasn't sure whether to rejoice or lament. Tonye and I had been married for about a dozen years. But life with her had become totally unbearable in the last five years or so. We have had no child between us, it is true, and I agree that to a large extent – as you yourself acknowledged in one of your letters without actually saying so – I agree that the absence of children has been responsible for the stresses in our marriage. But why must a man suffer in his own house simply because he has not given his wife a child? For some three years now she has not cooked me a single meal or even allowed me to make love to her – what have I done? She can't say I ever lifted a finger on her, despite all the provocation that could have driven me to it. I know I am an impatient man and I throw tantrums now and then and probably issue a few threats, but that's only because I have a restless energy within me, which I regret to say she has tended to distort. I am not a violent man. I do not enjoy doing harm to people and I abhor any sort of wanton damage whether to people's feelings or to things. So why should she torment me the way she has done all these years? Yes, I was happy I had married a Beniotu girl, and indeed my Beniotu nationalism was in many ways responsible for the tolerance with which I had borne my burden of provocations in more recent times. But it had become increasingly clear to me that she was more of an impediment than an aid to that nationalism. What use was it having a Beniotu woman beside me when I couldn't discuss our common ethnic interests and our people's goals? If I had a Diala or an Ibile woman as my wife, at least I would know where the lines should be drawn in our discussion and I could carry on my crusade without any fear or doubts about a conflict of interests.

In the end I was certain that Tonye's departure from my life wasn't such a terrible tragedy after all. I rose from the living-room and took a cab to Lati's place.

Let me not bore you or depress myself any further, Tonwe. It's been such a long letter anyway. I hope that when Priboye returns to Lagos he will come with some mail from you. I can't wait to hear the outcome of your visit to Benin. On my own part, I will not be satisfied until I have made contact with Bickerbug. I want to cool down for a while and recover from my encounter with our uniformed friends. In the next couple of days I hope to see things and perhaps crack the wall between me and the police. Meanwhile, please give my regards to your family. When I'm more settled I'd like Boboango to spend some of his holidays with me.

Very sincerely,
Piriye.

Lagos, 1.11.77

Tonwe,

I've been burgled! I swear, I've been burgled. I don't know who did it, but I have a good idea who they might be. And they're no petty thieves. What bothers me most is that I cannot reckon exactly how much they've taken away.

My wall clock says 6.28 p.m. I came in just over an hour ago from Lati's place. She was off work today, and I myself had decided to suspend by hunt for Bickerbug and other leads and relax a little. So I went over to her place about nine this morning. I had breakfast there, and lunch. We talked about a few things, and cracked a few jokes, but mostly she just let me rest – I particularly enjoyed the long siesta I had after lunch. Lati brought in most of today's dailies, and I had the rare leisure of reading through all the features and other stuff of a regular weekday. I even diverted myself with some of her women's magazines – *Woman's Own*, *Woman's World*, *Black Woman*, that kind of stuff. We chatted about my experience at the NSS, about Bickerbug, about you (nothing about our correspondence though), and a few other matters. Her aunt came in briefly and kept us company. Lati told her about my visit to the NSS and she sympathised and promised to talk to somebody – some lawyer or army officer, I can't remember now – so the police wouldn't harass me again. At about 5 o'clock or a little after that I told Lati I had to go, and she saw me off in a taxi.

I didn't notice anything unusual about my door as I was turning the key in the keyhole. Even now it's all perfectly normal – no damage of any kind done. The living-room itself looked reasonably normal. But I started suspecting something was wrong when I noticed that the glass doors of the bookshelf in there had been left open. I couldn't remember opening them myself before I left this morning. Also, on moving closer I found that the drawers at the bottom left side of the bookshelf had been pulled out and were in disarray. As I looked round, I could see a few papers lying on the floor near the door into by bedroom. Instinctively I moved towards the bedroom.

From the door I could see that the whole room was in a mess. One of my suitcases had been emptied onto the bed, the clothes thrown about on the bed and the floor. Another suitcase was on the floor, also with its contents turned out. I moved into the room and rummaged through the disorder and the items lying about. It was at this point that it dawned on me that I had not been visited by the common sort of thief we have all over Lagos. One of the suitcases had money in it – I can't remember how much I must have left in there, certainly not much more than N100. It all seemed to me pretty much intact. On further search, however, I discovered that my passport was no longer where it used to be, in one of the suitcases.

I moved from the suitcases and looked around. I had a briefcase where I kept copies of all the features I've written over the years. I looked for it in the corner of the wardrobe where it used to be, and it was gone. I discovered a few minutes later that the papers lying on the floor near the bedroom door were from that collection – they must have dropped out as my visitors were hurriedly leaving the bedroom.

You can imagine how disturbed I feel right now. One thing, though, set me at ease a little. Your letters are intact where they've always been. They were the next thing I went looking for the moment I discovered that the briefcase had disappeared. I'm surprised my guests never got to the place where I'd put them. They are contained in a file which I'd put in one of the drawers in a chest near my bed, at the bottom of a pile of odds and ends. The burglars didn't seem to have looked there. I suppose they thought the briefcase with the feature articles was a prize find, and didn't think about looking elsewhere once they'd got that. I heaved a sigh of relief when I found the file was still there. But I can't tell what else they took.

Anyhow, I'm not taking any chances. I feel quite certain I've been visited by the NSS. I don't know how they could have come in so neatly without breaking the door. Could my wife – I mean, ex-wife – have a hand in all this, or did my visitors come in with a master key? Well, I can't imagine what else might interest them here henceforth. But to be on the safe side, I'm putting all your letters in a big envelope and taking them for safe keeping to Lati's place. Be assured I won't tell her what the contents are – I'll simply tell her they are personal papers.

I hope Priboye comes soon. You'll hear more as matters develop. Meanwhile, please be careful yourself. I don't know how much our friends know. Or will want to know. God damn it!

What's this country coming to?

Piriye

Brisibe Compound
Seiama.

28 October, 1977

Dear Piriye:

I will tell you about my trip to Benin. But first you must excuse a little digression. It has been a couple of years since I left the *National Chronicle*, and I must say I cannot quite remember when, in all the years of my service there, I handled an investigative assignment with the sort of disposition I find myself bringing to the project in hand. In my time, as you no doubt know, I did a few such chores across the country: the customs racket at the Nigeria-Niger border, illegal mining in the plateaux of the Middle Belt, labour riots at the Railway Corporation in Lagos, and a few other similarly sensitive assignments. When I did those assignments, I was a much younger man. Perhaps I should say that, in those days, a combination of youthful exuberance and professional zeal caused me to be easily attracted to the more practical details of human behaviour, in my anxiety to turn out stories that sizzled with the heat of action. I wanted to beat my competitors on the other newspapers to the press. Seldom did I settle down to explore deeply the motivations behind the actions that I reported. At any rate, I was too busy jotting down notes to take a close critical look at the faces and postures of the subjects I was interviewing.

In the present case, however, I find I am allowing myself considerably more time and caution to observe things: the angle of a smile, the movement of the hand, the time it takes a man to ponder a question, even the pace of his breathing. You might say this is all because I am no longer under the sort of pressure that I felt as a young man struggling my way up the ladder at the *Chronicle*. I might also concede that growing older has a lot to do with this new disposition. But I think also that coming home to the village, living and mixing with the simple, honest folk of our fishing communities and encountering life, as it were, at its most basic level, has disposed me readily now to seek a more intuitive grasp of things, than I could have done in the frenzied pace of professional life in Lagos.

Mind you, I do not like what I am beginning to see. Indeed, it is only with some effort that I am managing to keep in check two conflicting sensations that have lately grown within me. At one time I feel grateful to you that, in participating in this investigative project, I have been given an opportunity to distil my accumulated wisdom into a document that may benefit our nation, if only our leaders will listen to the message of it. At another time I feel a little resentful at being torn from my well-earned (though forced) retirement into a theatre of action that, from what I can see at this point, promises more turmoil and more storms than I ever rode in my professional youth. But I have resolved to give this project as much time as my family commitments will allow and as much energy as I can still muster at my age, not only because I now bear the charge of a community that seems otherwise without a hope but also because, shielded at least by the collective empathy of my kith and kin, I am moved to register my own testimony against evil and injustice in this country.

Let me stress, however, that I will continue to avoid the confrontational approach that the likes of Mr Harrison seem to favour. I see nothing in it but trouble for our people. And nobody knows how far beyond the Delta the repercussions may go. Not so long ago this country survived a civil war which started when a small group of army officers eliminated an equally small handful of politicians who they thought were the only obstacles to peace in the land. No one who lived through the trauma of that history would in his wisdom support the use of violence. And my experience with Navy Commander Adetunji in Warri persuaded me to avoid taking along with me anything or anybody whose presence might cause any provocation to the authorities whose cooperation I might hope to enlist in this cause. Which was why I went to Benin alone, and without the knowledge of anybody but my wife.

In Benin I was the guest of Mr Johnson Aniemeka, Permanent Secretary at the state's Ministry of Information and Culture, an old friend from our days as pioneer students at Ibadan in 1948. He welcomed me warmly. He was glad to see me back in the metropolis. The last time I saw him was about two years ago, on my way home to my village after the forced retirement. I remember him trying to convince me that I was making a mistake in surrendering my talents to the dull life of the village, when there were a thousand things I could busy myself with in Lagos.

He said I was one of the finest journalists this country had ever had, that talents like mine were still needed, no matter how much wrong I had suffered. Of course, he had no way of knowing just how much wrong I had suffered. I was deeply grieved then. It took me some time after settling down in my little village to overcome the very strong feelings I had. But I was glad I did not lose my temper all through our discussion that afternoon, two years ago. I knew he meant well, although he had no understanding of how badly I had been hurt at the *Chronicle*. Now, two years later, when I told him how contented I was with life in the village from which I was managing to eke out an adequate existence, he seemed to resign himself to my outlook on things.

I did not tell Aniemeka the real cause of my visit to Benin until we had had supper. I did not tell him anything about the investigative project. I told him about the rumblings down in the Delta, about the threat to our sources of livelihood, about the experience of Opene and the fishermen at the rig. When I then told him about my visit to Commander Adetunji, I noticed he leaned back in his chair and regarded me with an added curiosity. His posture at this point rekindled in me the feeling which I had after I made that ill-starred visit and which I think I must have conveyed to some extent in my last letter to you: am I not bringing into this issue a twist and a complication which will remove it from the level at which it has been safely played so far? Am I not therefore liable to do more harm than good to our people by playing an overzealous champion of their cause?

Now, I am still not sure if Aniemeka's quizzical look was from a genuine concern at the dangers he feared I faced or if, suddenly remembering that he was a civil servant and I a journalist (although a retired one), he was now viewing me with the suspicion of one who feared I was subtly probing the wall protecting his official secrets. Up to this point he had talked quite liberally, but now he was doing more listening than talking. When he did talk, it was in more or less non-committal terms.

'So what do you intend to do about the problem?' he asked.

'Good question,' I said. 'I am not even sure I understand the problem correctly. We are not asking that oil exploration should be completely abandoned. Nobody who has any interest in the development of the country would contemplate such a request. But is it so difficult for the government to concede that it has a duty to guarantee the lives of the citizens of this country, and the sources of their livelihood? That cannot be too much to ask.'

'I agree with you,' said Aniemeka. 'I agree with you.'

For a while we looked at each other. I wished he would say something more helpful. But he seemed unwilling to commit himself any further than he had already done.

'Now,' I said, 'it is clear to me that the poor fishermen in the Delta cannot do anything to help themselves. Those of them I have spoken to so far are decent people. They don't want any trouble. There have been outbreaks of violence now and then. But these are really the handiwork of troublemakers in the area – mostly jobless or irresponsible youths coming into the area from the cities just to make trouble. I was hoping that I could talk to a few people here at state level and so help to keep things under control. I am aware of certain moves being made in Lagos. But I have my doubts about their usefulness. Surely, there is something the state government can do – after all, it's in charge of this area. I believe Iyamabo is still Commissioner for Health and Environmental Affairs?'

'Iyamabo?' he laughed. 'Good God, Brisibe, where have you been? There was a cabinet reshuffle way back last year, and Iyamabo is out of office. Are you so out of touch?'

'I'm afraid so,' I replied. 'So who replaced him?'

'Batowei.'

'Who?'

'Batowei. Freeborn Batowei. You know him?"

'Yes,' I said. 'I do, I know him.'

Do I know Batowei! We were classmates all through primary school in Burutu, and we were very close. When I left and went to secondary school in Lagos, he stayed on in Burutu to work for the UAC there. Years later he went off to read law in England. After his return he practised law in Warri, not very successfully, I'm afraid. He tried politics, but lost woefully in the elections. You must have read about him. Then he went into business, again in Warri, and has done quite well for himself. I am told he has the best looking house in his home town, Brofani. I did not realise he had gone into government. But anything can happen these days. He is a most likeable fellow, very open-hearted. I remember him too as a hardworking and conscientious person. I am not surprised that he keeps rebounding from his setbacks.

'Yes, I know Batowei.'

'Well, if you know Batowei, your problems are solved,' said Aniemeka. 'He's the man you should talk to. Besides, he's from your area.'

I nodded. Aniemeka's face brightened once again. He seemed to be happy to have the responsibility for giving information shifted from him to someone else. Which was all right by me. I was glad to have run into the good luck of finding that the man in charge of the business that brought me to Benin was an old childhood friend. And since he came from the same area of the state as I did, I was sure he would have a greater feeling for the problems we faced there than someone from a different area. I was glad of the information, and would be happy to rid Aniemeka of the embarrassment of discussing matters he would rather not volunteer information on, even if they were within his area of responsibility.

Aniemeka has done his best, I must confess. Let us be honest: what else could I expect of the man? If a man like Opene, a poor illiterate fisherman who knows little about the intricacies of government policy and the international manoeuvres surrounding the petroleum industry, knows enough to tell that oil is trouble, you can imagine how deeper an understanding (not to say inside knowledge) is available to the likes of Aniemeka who help to shape that policy. And would it be fair of me to expect him to put his career in jeopardy, to endanger the livelihood of himself and his family, just to satisfy the curiosity or the zeal of a man who has elected to seek a safe retreat from these problems by retiring to his village? On pondering Aniemeka's behaviour, I was satisfied and grateful simply to have been given the information that would take me one step further in my investigations. I think I slept well that night.

The following morning he very kindly made an appointment with Batowei's personal assistant, for me to see the Commissioner. Later that same morning, at about 11.30 a.m. when I was due to see Batowei, Aniemeka told his driver to drop me at the Ministry of Health and Environmental Affairs. I thanked him very much indeed for his kindness.

You know what it is like seeing a Commissioner. I had to wait in a small receptions room, because Batowei was having a meeting with one of his committees. I was there for about an hour. His secretary very kindly served me coffee and crackers. I even broke my embargo on newspapers and amused myself with the morning's papers, including our old friend the *Chronicle!* I was pleasantly surprised to find that the editor was now Mr Ojulari, that fine young man I had brought in from the *Standard.* I glanced through a few features. I felt quite comfortable, not having been in an air conditioned office for two years! While I was leafing through the papers, the door opened and there was Batowei, having finished his meeting, exploding on me with a loud greeting and a bear hug. He led me straight into his office and shut the door. Safely inside, we escaped into the past and, like little children, regaled ourselves with an old comradely chant and the appropriate bodily movements!

It was a wonderful feeling, I can tell you. Of course, since coming home to the village, I have been reunited with our folk ways and traditions. But doing that chant with Batowei was like being back in the Burutu of my early youth. You can't imagine how much joy and empathy we felt at regaining each other's company and being youthful again. I had always liked Batowei a lot. That brief moment helped in no small way to set our meeting on a sound footing of fellow-feeling. We were full of smiles.

He ushered me into a very posh armchair in a corner of his large, cool office, and took his seat in the chair nearest mine. He was going to call for his secretary to serve us coffee, but I told him I had already had some refreshment.

'Well,' he said, surveying me with laughing eyes, 'it's been a long time, Tonwe, eh?'

'Very long indeed,' I replied. 'Some forty years perhaps. Isn't that a terrible thing?'

'How come we haven't seen each other all this time?'

'Well,' I said, 'you know what it's like. We have both been very busy people. I've been in Lagos, at least until not so long ago, and you've been in Warri and Benin. But I've been following your progress and hearing about you from our mutual friends: Youdeowei, Ikime, Ukoli, Sogolo, Okudu, Ekpere, Dede. I'm very happy for you.'

'We thank God,' he smiled shyly. 'Yes, I too have been following the news about you. I'm sorry about the terrible deal you got from the *National Chronicle.* Everybody was quite scandalised.'

'That's life!' I said, with a strained smile, hoping he would realise I wished he would drop the subject.

'Well,' he said, getting the message, 'you look very well. Going to the village has obviously not been such a bad idea.'

'Not really. We manage to survive. By the way, congratulations on your house in the village. I hear it is quite a magnificent place.'

'Thank you,' he smiled again. 'I've only recently finished it. In fact, I moved in only two weeks ago. Er ...'

I think he realised he had said something he should not have said, because there was a touch of disorientation on his face. The intercom on his telephone rang, and he sprang up quickly to answer it, excusing himself as he rose.

He picked up the receiver and pressed the button.

'Yes?'

His secretary was obviously telling him about some newly arrived visitors, and took some time introducing the subject.

'Do they have their forms with them?' Batowei asked.

The other voice spoke. Batowei's face became fairly serious.

'Tell them to bring their forms, properly filled in, and return tomorrow morning about this time. All right?'

After a brief hesitation, he dropped the receiver. The smile returned to his face as he walked back to his seat near me. But it was no longer a very warm smile. It was now a polite smile, a courteous smile.

'Well, Tonwe,' he said, 'what has brought you to Benin?'

'Ehm,' I said, adjusting my posture and clearing my throat, 'it's a problem very much on your doorstep. I have come to you first because we are friends and then because, if there is anything that can be done about it, you would at least be willing to give us your support.' He was now sitting on the edge of his chair, his mouth half open and his forehead visibly furrowed with anticipation. I went on. 'It is about oil in the Delta, and the difficulties we face from it. It is about fishing, and our whole livelihood, and how we can no longer be sure of the future.'

I thought I had made my purpose clear enough. I waited for him to at least indicate he followed me. He nodded a few times. He looked at his watch. I think it was a little past 1 p.m. He cleared his throat.

'Tonwe,' he said, not looking at me now, 'what you have brought up is a very important matter. But I wouldn't like us to discuss it here. By the way, where are you staying in Benin?'

'With Johnson Aniemeka. We've known each other since our Ibadan days.'

'Of course,' he said. 'Still, you don't mind having lunch at my place, do you? Beside, I'd like you to meet my family.'

I said it was a great idea.

'Good,' he said, brightening up again.

He rose quickly, pressed the intercom, and told his secretary he was leaving for the day. He called for his driver. When the young man came in, Batowei handed him a briefcase to take to the car. Then he put a few bits and pieces in his pockets, gave his table and his room one final check, and said, 'Tonwe, let's go'. I rose and followed him. We went outside to the official-looking car waiting at the entrance to the building. A policeman saluted and opened a door. I went round to the other side and sat beside Batowei in the back.

'Do you come to Benin often?' he asked, after we had left the Ministry's premises.

'No,' I replied. 'In fact, this is my first visit here in about two years.'

'Really? How come?'

'Trying to settle down in the village,' I told him. 'It has not been easy adjusting to life there. But I had made up my mind I was going home. I knew things would be rough for a while. Anyway, we are managing to survive.'

He shook his head. For some time we did not speak. Once I looked briefly in his direction. From the way his shoulders were squared I could feel the force of his official personality setting a distance between us. Perhaps it was the well-cut suit he was wearing and the simple Beniotu tunic and trousers I had on. But partly also it was the serious look he now wore on his face. I had a feeling I had set him a problem. The drive was probably a good opportunity for him to turn it over in his mind and prepare himself to respond. He did not seem to notice me when I looked at him. I thought the silence was not good for us. I never like making people unhappy. So I decided to break the silence.

'Tell me about your family before I meet them,' I said.

'We have four children, all boys. One read Medicine in Lagos and is doing his house job now. Two are at the University of Benin. The last one is in secondary school here in Benin. My wife is from Ogwashi-Uku,' he said, looking at me with an apologetic smile.

'Very good,' I reassured him. 'We are all the same.'

'How about you?'

'My wife and I have two boys and a girl. The first boy read Medicine at Ibadan and works at the Teaching Hospital there. The girl read Classics, also at Ibadan, and got married last year. She and her husband are in Warri. Our last child is at Urhobo College in Warri. Only my wife and I live in the village. Incidentally,' it was now my turn to smile apologetically, 'she also comes from Seiama.'

'Oh,' Batowei said, 'lucky man.'

'Not really,' I said. 'It doesn't matter, I think. We just happen to have grown up together.'

Batowei lives in one of those big colonial residences in the GRA, quite close to the Military Governor's premises. A huge compound he has there, with well-manicured lawns. The roof, which is high, is made of asbestos plates arranged in terraces. The car stops, and the police guard in the compound runs up quickly to open Batowei's door. We get out of the car and walk up the steps that allow a good clearance between the ground and the living room. A steward opens the front door for us to walk in, and collects Batowei's briefcase from the driver, who has come up behind us.

'Momodu,' calls Batowei. The steward answers and runs up to him. 'Set the table for me and my friend, and make more food for madam and Ditimi.'

'Yes, sir,' says the steward.

'My wife has a hair-dressing salon on Mission Road,' Batowei tells me, as the steward disappears into the kitchen. 'She doesn't come home until about 6 p.m. My son Ditimi is a day student at the ICC. The driver will bring him home about 2.30.'

'Right,' I say. 'Nice place you have here.'

'Not bad,' he smiles. 'I imagine I'll be here until the next cabinet reshuffle, or the next change of government.'

We laugh over the joke, though we know either of these is a distinct possibility. The steward turns on the air-conditioner as we step into the living room. The cool air is a great relief from the heat outside. The steward reappears to ask what I will drink, and I say Star beer, which he serves me. I am genuinely impressed by the good taste shown in the choice of living room furniture, and the colour scheme of the blinds, the chairs and the rug. In this rather relaxed mood, we reminisce some more about old times. As I indicated above, this is my first meeting with Batowei in about forty years. After I left Burutu for Lagos I never saw him again until this afternoon. So we recall old names, old experiences, old pranks, and discuss other things: the slowly rising prices, attempted coups, the military governor and rumours about the Head of State moving the present set of governors around, agitations for the splitting of our state into two or three, and so on and so forth. The steward returns to tell us lunch is ready. We move to the table. It is a delicious meal of rice and fried plantain, washed down with more beer. Batowei and I crack more jokes and recall more names. As we rise from the dining room to return to the living room, Ditimi, aged about fourteen, returns from school. He is introduced by his father.

'Ditimi, which is your favourite newspaper?' Batowei asks.

The boy ponders a little and then says, '*The Standard*.'

His father and I laugh, though Batowei is slightly embarrassed.

'You mean you don't read the *Chronicle*?' he asks the boy.

'Sometimes, yes,' says the boy, looking at me now and realising he may have said the wrong thing, 'I like it too. I like the crossword puzzles and the cartoons.'

'Well,' says his father, pointing to me, 'here's the man who made the paper.'

The boy looks at me, half-delighted, half-awed.

'Never mind, Ditimi,' I reassure him. 'You are right in liking the *Standard*. It is a very good paper.'

I ruffle his hair, and he disappears upstairs. The clock on the wall says something like 3.15. Batowei and I settle down once again. By now the laughter has cooled down between us, for we both realise it is now time to get down to the business that has brought us from the office to his house. He takes one more draught from his glass, and clears his throat.

'Tonwe,' he says, 'I'd be a fool to pretend I don't understand the nature of the business that brings you to me. But tell me at least what you have in mind. What has happened? How can I help?'

I pull myself together and take a deep breath.

'Well,' I begin, 'after I came home from Lagos, I decided I had

had enough of the hustle and bustle of the city and should retire to a quiet rural life. But something happened about two months ago which convinced me that I had to get involved in the life around me, if I truly meant to earn my peace.'

Batowei is looking at me very intently, very serious, his chin resting on his interlocked fingers and his elbows on his knees. My tone is low, almost conspiratorial, although we are alone in the cool, quiet living room, effectively shut off from Ditimi who is eating in the dining room, and the steward further beyond. I tell him everything: about my meeting with the fishermen, their report about the incident at the oil rig, their appeal to me to take up their cause. I tell him about my long reflection on the matter, about the visit Opene and I made to the garrison commander in Warri, how we were treated there. Every once in a while he nods or grunts, but never changes his posture. I make him understand I had no idea who I was going to meet on my trip to Benin, but that I had resolved I was going to explore every possible avenue until at least our case was lodged somewhere where it stood a chance of being heard. I assure him I have come on a mission of peace. I am not ignorant of the agitations over the problems of the Delta, both in Lagos and in the Delta itself, the skirmishes at the oil installations between villagers and law enforcement officers, which have resulted in some damage and some casualties. I let him know that these unfortunate cases of violence are the main reason I have decided to make this quiet trip, in the hope that I can make a quiet case with someone who will appreciate the need for a peaceful approach. After I have finished speaking, he sighs deeply, disengages his hands, and rests his back more comfortably on the chair.

'Tell me, Tonwe,' says Batowei, 'what has been your own personal experience in the Delta? I mean, how much has this pollution from oil affected your own existence?'

I ponder the question for a brief moment, for I concede it is a good one.

'Well,' I finally answer, 'not as much as it has affected others. In a sense, my village is luckier than many in the area. The drilling operations are at least a good two kilometres away from us. Most of the land is protected by a network of fine matted growth that seems to keep the flotsam well away from the land. A clear lagoon recedes into the belly of the land almost like a lake, and its neck too is girded by some of those matted weeds, so that we have to sweep them with our paddles as we row into the open. We do most of our fishing within this area: in fact, we have the sole territorial rights to it, if I may say so. Yes, we are luckier than most villages, I think. But that does not mean we are entirely out of trouble's way. A few weeks ago there was a huge underground explosion about one kilometre offshore from our village. My wife and I thought it was an earthquake, but we were assured later that it was a fairly common thing in our area. Not long afterwards, some of the spillage floated very close to our banks.'

'Is that all?' asks Batowei.

'Not quite,' I say. 'The effect on our agriculture isn't really disastrous, but the yield is not so great either. And my wife has noticed something about our drinking water. Of course, you know how susceptible women are to these fears. But she has constantly told me, especially in more recent times, that the water tastes strange, as though there is some kerosene in it.'

'Where do you get your drinking water from?' asks Batowei.

'We have a well in our compound, which I dug myself, years ago. We used to boil and filter the water in the first two or three months after we came home from Lagos. But we abandoned the habit, which we had indulged largely for our son Boboango's sake. We reasoned that, if we were going to live the rest of our lives in the village, we might as well adjust ourselves to its problems. To be honest with you, I may have noticed the same taste in the water myself. But I think I am so determined to stay and survive in the village that I have not let that taste bother me. And you know what women are like. If I made the mistake of confirming her misgivings, she would probably begin to suggest that we move back to the city, just to save our health!'

We both laugh at the joke. But Batowei is a little less amused than I am. In fact, I can see a slight restlessness in his posture. He tightens his fingers now and again round the tips of the arm-rests of his chair, shakes his legs ceaselessly. The laughter soon disappears from his face.

'You know, Tonwe,' he says, 'I'm glad you made that point about your determination to settle at home in the village. You see, we are not getting any younger.'

I nod in agreement, though I begin to wonder which way he is leading.

'If you'll take my advice,' he continues, 'please don't get yourself involved in anything that may ruin your chances of realising this happy life in the village.'

'I am not sure I understand,' I tell him.

'Let me be quite frank with you, Tonwe,' he says, 'because you and I have known each other too long for us to be talking around each other. It was exactly for this reason that I decided to bring you to my house rather than talk with you in my office.'

The telephone rings. He excuses himself and rises to take the call. From the discussion, it looks like a call from his office. It cannot have been a matter of great importance, for he soon disposes of it and returns to his seat.

'Take a look at me, Tonwe,' Batowei continues. 'I've come a long way. Like you, I started off as a poor little country boy, son of a fisherman. I've had my share of suffering. You know how hard I've had to work all my life, right from when we finished primary school in Burutu. All the ups and downs, and more suffering. You know all about that.'

I nod again. Without wishing to encourage any self-pity, I must acknowledge that the fellow has paid his dues.

'And now,' he continues, 'you think I'm going to let anyone deny me my well-earned rewards?' I shake my head. 'That's why I asked you, how has the oil problem affected *your* life? I really meant to ask you what's in it for you?'

'For me?' I ask him, almost in disbelief. 'Personally?'

'Well,' he says, 'I know how you feel about this. I've always known you as a principled and fair-minded man, right from our primary school days. And I know that you wouldn't come all the way here from the village, if you didn't feel strongly about this matter. But look at the country you're fighting to save. Is it really worth all this effort? We fought a civil war. The war has only been over seven years or so, but where are the people who fought so hard to split the country? They are today holding some of the topmost official positions in the country. Even in the *army*. Even in *petroleum*, which you are so bothered about: where do you think the Federal Minister for Petroleum comes from?'

'He is a Diala man,' I reply.

'There you are! All right. Look what a mess we're in in this country right now. Every other year, there's a coup, and a new government comes into power. New men, new policies, new budgets, new spending – and we start all over again. Not that the civilians would do any better – they're probably worse, if you ask me. But this constant change leaves us without anything to hold on to, anything to believe in for any length of time. You have come to me with a problem, and perhaps if we got down to it we could knock together something that might sort out the problem somehow. But what's the use? Tomorrow a new governor will be brought to this state, and he might replace me. If you're still interested in your problem, and the new commissioner is interested, he will want to revise everything we've said and done. When you think of all this, Tonwe, you'll find there's no use fighting for a cause. These are not times for idealism, my friend. Get what you can before it's too late.'

I have been looking at him and nodding all this time. He does not look particularly taxed at the end of his moving rhetoric. Although he takes off his jacket and his shoes and loosens his tie, he is obviously still in tune for further rhetoric if he gets the necessary encouragement. From the way he returns my stare, I can see he at least expects that an old friend will reciprocate his frankness and good faith.

'I agree with you, Freeborn,' I tell him. 'There is a lot that is wrong today. I appreciate what you say about my being a principled man, but I am not so naive that I do not know what is going on in this country. Perhaps I need not remind you that I am home on retirement today, precisely because of what is happening in the country. You ask how this oil business has affected my life and what is in it for me. Let me put it this way. I have come home for a peaceful retirement. But if things continue the way they are going, there may be nothing peaceful about it. I have been driven from Lagos. Now I am home, there is nowhere else for me to run to. Take your mansion in the village. I imagine you have built it so that, when you finally retire, you will have a good place to go home to and spend the rest of your days in peace and comfort. Now, how would you like it if this whole oil palaver got out of control, and your own village, and with it your beautiful house, was bogged down by oil spillage? So, if you ask me what is in it for me, I say it comes down to that.

'But let us look at it from another angle. We talked about the civil war, and we should always remember that such things usually start from one small localised problem, which ends up engulfing the whole country because nobody is really interested in giving a chance to a peaceful approach. So the fishermen get angry and blow up all the oil installations, and the Federal Government brings down its might on our people, and another Isaac Boro rises to try to fight the people's cause, and the whole Delta goes up in flames once again. Now, where does that leave you and me, and all the beautiful homes we hope to retire peacefully to? My dear Freeborn, I am not being idealistic. I think I am being as realistic as any man who has seen enough of this country ought to be.'

We must have been speaking at the tops of our voices, and leaving the impression that we have been having an intractable quarrel. Ditimi has been having his lunch in the dining room. Every once in a while he has peered into the living room to see what has been going on between his father and me. On finishing his meal, he worms his way into the living room. He perches on the arm of one of the sofas, obviously anticipating a reaction from his father. He is not disappointed.

'Ditimi,' says his father, not particularly hostile, but not too cheerful either. 'Have you finished your lunch?'

The boy nods, rising up from his seat.

'Why don't you go up to your room?' says Batowei. 'We are discussing a rather important matter.'

'Let him stay, Freeborn,' I intercede. 'He is not disturbing us. Besides, he may learn a thing or two.'

'No,' replies Batowei, firm but not hostile. 'Let him go up and have a rest.'

The boy sighs, excuses himself, and walks upstairs.

'Freeborn,' I say, 'you should have let him stay to listen. He may be a young boy now, but he won't be for ever. And the young are the ones we should really be fighting for, not ourselves. We are on our way out, and I don't think we have a right to leave the homeland in ruins for generations ahead who may want to make such a home for themselves here as you and I are busy doing for ourselves now. That's really the point I have come to make to you today.'

Batowei shakes his head and sighs.

'You've got it wrong, Tonwe. And that was the whole point of my asking you what's in all this for you, and urging you to take what you can for yourself before it's too late. What makes you so sure Ditimi and his generation will be interested in building a home in the village, whether now or in the future? Times have changed, my friend. True, I've built myself a mansion in the village, and I hope to retire there. But that's because I remember the village. I made my roots there before I left it. But what does Ditimi know about our village? Nothing, practically nothing. I make the ritual of taking my family home about twice a year, so at least my children may know where they come from and who their relatives and their people are. My wife also takes them to Ogwashi-Uku once in a while, and I don't mind. But I'm not so sure they feel about the village as deeply as I do, because they haven't grown up there. They seem to feel much closer to Warri, and perhaps Benin now, than they do to Brofani. Many a time when I've taken them to the village during my leave, scarcely would a week pass before they would say to me, "Daddy, let's go home", meaning Warri or Benin.'

'I know what you mean,' I say. 'That also happened to me. But they grow up and realise that their real home is in the village, not Warri or Benin.'

'I hope so, but I'm not so sure about that. Nigeria is changing fast. We're talking about oil. There are families from all over Nigeria now living in villages like Escravos and Burutu, or even cities like Warri and Port Harcourt, who have come to make their fortunes working for oil companies. Many of them hardly ever go home, for various reasons. Tomorrow, where do you think their children will feel their homes are: the ones their parents come from, which they have never seen? The same thing may happen to your own village, Seiama. What makes you so sure that, twenty years from now, with the expansion in oil exploration, the non-natives living in Seiama will not outnumber the sons of the land there? So what are you fighting for?'

'The land,' I say. 'The land and the people. Whoever makes his home with us is one of us. What does it matter where he came from? Our children and relatives are living there where he came from. So what does it matter? Does that make our duty to save the land, that protects and nourishes them and us, any less urgent? No, it doesn't. And that's what I'm fighting for.'

It is clear we are not making much progress in our discussion, because we are approaching the issue from two different viewpoints. It is no longer clear to me which of us is being idealistic and which realistic. But anyone who can sit there and tell me that it is more in my interest to make money, while I watch the oil corporations and their collaborators inside or outside government destroy our environment and our lives, is certainly not being any more realistic than myself. Batowei is contemplating the floor, and at this point I am not so sure what to say next. The impasse is becoming increasingly awkward, when the door opens and in steps a very large woman.

We both stand up in obvious relief, and Batowei introduces his wife to me. We try to be at ease, and I must say she is a pleasant lady. She has not come home to stay: she has simply forgotten something and has come to pick it up. She tries to persuade me to spend the night, but I regret I really cannot. After a few more pleasantries, she excuses herself, goes upstairs, and in a short while is back again. She leaves a few instructions for the steward, then apologises profusely to me that she cannot stay with us. We bid each other goodbye. I promise I will revisit some other time. She goes out and is driven away by the driver.

Piriye, I am sorry that this report has turned out to be as long as this. Believe me when I say that I have been writing it for three days, in between my regular chores. I am a little tired. But I thought the experiences I went through, and the reflections they forced upon me, sufficiently significant for me to devote some time to spelling them out in some detail. Besides, Priboye will be going to Lagos soon, and I do not know when he will come this way again. Anyway, let me conclude the report on my visit with Batowei, before I close this letter.

After his wife left, we did not spend much longer on our discussion: partly, I think, because the momentum of the debate had been broken by her entry, but partly also because our lines of approach to the problem did not hold much promise of meeting. It seemed to be also, on a closer look at the man at this point, that the interlude of his wife's appearance had given him an opportunity to reconsider his openness to me. He had brought me home to his house to discuss a delicate official matter. He did so because I had touched his heart with memories of our boyhood days, and he had felt disposed to bare his heart to me in his house, rather than speak guardedly with me in his office. Now that it was clear we did not share the same sentiments on the subject, he seemed to have gradually resumed his official distance. His face had become somewhat sterner. He now seemed more inclined to offer advice and state the government's position on the matter than to discuss the issues involved.

I will not bore you with more details of the discussion at this point. He said the usual things you would expect of a public servant: the value of oil to the economy, the oil pollution as part of the price we have to pay, the government's deep concern for the welfare of the people most immediately affected by the hazards, and so on and so forth. When I asked him in what practical ways the government was going to demonstrate this concern, he told me something which provided a good opportunity for us to bid goodbye to each other.

'The Ministry of Petroleum and Power has set up a committee to investigate the environmental and developmental problems of the oil-producing areas, and make concrete recommendations. Members of the committee will be undertaking a tour of the areas soon. My ministry has been asked to coordinate the input of the state government to their efforts. We are putting together a team of our own to visit the Delta with the men from Lagos. If you are interested, I will appoint you one of our key consultants. The team will visit you or may require that you do some work with them. Are you interested?'

'I don't mind,' I said. 'But what kind of work would they want me to do with them?'

'I don't know,' he said, his face now impassive. 'When they come they will let you know their terms, and I suppose everything else will proceed from there.'

For a while we looked each other in the eye. Then I said, 'All right', and rose to go. The pleasantries of parting were not as warm as those of our meeting some hours earlier. But we smiled nevertheless, and promised to keep in touch. The driver took me back to Aniemeka's house. The following day I left Benin for the Delta.

Two things bothered me about my visit to Benin. I am no stranger, of course, to the ways of public officials. It is true that my quiet and more traditional life-style in the village has begun to remove me from the sort of scenes and atmosphere I regularly encountered as a journalist in Lagos. But I had never felt so sorry for this country as I did at that moment. Maybe Batowei was right: living a quiet life in the village has probably started to shield me from the harsh, impersonal realities of Nigerian life and to make an idealist out of me. But I had never felt so sad as I did at that moment, watching Batowei. In only a few hours, he had changed from a regular bureaucrat, flanked by telephones, secretaries and drivers who did his will; to a heartwarming home-boy exchanging reminiscences of our mutual past and confiding honest sentiments to me; and back again to the impassive, straight-faced bureaucrat enunciating government decisions to me, now and again perhaps with a mischievious gleam in his eye.

I could see the catch in that invitation to join the investigative team touring the Delta. Batowei was convinced, beyond any shadow of doubt, that I was not standing on his side of the line. Rather than that I should constitute an obstacle to his game of survival, he hoped that the investigative committee, which I am sure will be no more honest than several such bodies we have seen set up in this country, would give me enough reason to change my line of thinking on these matters. As I observed these various changes of posture in the man, I was not so certain which posed the greater threat to the survival of our people: the shameless duplicity of the bureaucrat, or the bare-faced insolence of the braggart soldier.

That phrase "our people" leads me to the second revelation that dawned on me as I pondered my experience with Batowei on my way back to the Delta. The demographic analysis which he made during our discussion forced me, quite frankly, to take a fresh look at the nature of my commitment to the cause which I have undertaken to champion. I recall urging you, earlier in our correspondence, to avoid taking a narrow ethnocentric view of the troubles facing this country, and I have seen no reason yet to alter my position on that matter. But what kind of "people" do I really have in mind, especially in relation to the crisis in the Delta? Up to this point I had unreflectingly assumed a monolithic Beniotu ethnicity when thinking, for instance, of my own little village or the cluster of communities in our area. I have always seen Opene as one of us, even though he is not strictly a Beniotu man, because he has lived long with us and shares our deepest sentiments in these troubled times. And I am glad I expressed the same conviction to Batowei during our arguments.

But, suppose someone else in these troubled creeks, who is not Beniotu but lives among us, were to behave in a manner contrary to Opene and set himself stoutly against our well-being: would I continue to uphold this campaign against the demographic realities Batowei has outlined to me so starkly? In saying that Opene is "one of us", am I not treating him as a Beniotu man and thus assuming that the cause of this community which I am championing is necessarily a Beniotu cause and thus, by extension, that only the Beniotu have a natural proprietorship over this area in which we happen to be living?

Forgive me if I seem to be making this issue more complex than it already is. I simply want to be sure that I am fighting this cause with the right sentiments, and that I am bringing these along in cooperating with you in this investigative project. I forgive the Batoweis and the Adetunjis of this country, of this world, because we live in a complex human family and, in the end, it is no longer easy to draw a line between what is idealistic and what realistic. All I can say is, may God direct our efforts in this cause.

I am waiting to see what the investigative team from Lagos will be coming here to do. You will hear further from me about any developments in that direction.

I must go off now to inspect my fishtraps. My wife joins me in wishing you the best of health. She insists on adding her anxiety to know how matters stand between you and Tonye. Please take good care of yourself.

Yours sincerely,
Tonwe Brisibe.

Lagos, 12.11.77

Dear Tonwe,

Priboye's visit was timely. I don't know whether to describe it as a relief or something else. He came in about fifteen minutes after I returned to my house from my detention by the NSS. I leave him to tell you in what state of mind he found me.

I'm not surprised at the experience you had with that asshole Batowei. Every ethnic group has its misfits and traitors, and the Beniotu have had their fair share of the Freeborn Batoweis, the Brown Siekpes and the rest of them. But let me postpone further comment on him and the proposed visit of the investigative team until I've told you what happened to me within the last few days.

Although I was shaken by the removal of my papers by men from the NSS and hadn't quite recovered from the shock, I recalled I had two deadlines to meet with the *Standard* and *Ebony* magazine and simply had to settle down and do those stories. Lati had offered that I could do anything I wanted at her place, but it's not quite the same thing. You know what I mean – you know where to look when you need a particular book for reference, you want the freedom to get up any time and grab a snack or put on a particular kind of music when you get bored, or pick up a magazine story or a book where you left off reading it. That kind of thing, you know.

So four days ago I was sitting at my writing table putting together this feature on environmental sanitation for the *Standard* when I heard a knock on my door. About 5 o'clock, I think. I wasn't expecting any visitor as such, although Lati could have come any time. But that wasn't Lati's way of knocking, and I thought perhaps Priboye was back. I got up from the chair and opened the door. Two men in French safaris came in, and as one of them said to me, 'You are Mr Dukumo, aren't you?', the other, stout, bearded and grim-looking, elbowed his way brusquely past me and said, 'You know he is, Osawe. Let's cut the politeness and get on with our business.' He gave me a hard push on the chest which nearly had me over and said, 'Sit down, Piriye. We want to ask you a few questions and please don't give us any bullshit. We haven't got any time to ...'

'Come on, Dayo,' his companion said, holding him by the arm. 'Take it easy, man, let's not push him about yet, not until he refuses to cooperate with us.' And turning to me, he said, 'I'm sorry, Mr Dukumo. We're really not like that. But you don't mind if we ask you a few questions?'

His companion didn't appreciate that style at all and walked away to inspect my apartment. They weren't welcome – no sir! – but I motioned Osawe to sit down while I did the same, turning to cast a hateful glance at the wild one.

'I'm Phil Osawe, and my companion over there is Dayo Haastrup.' He put a black briefcase down beside him.

I simply nodded. What the hell – they already knew my name, so what was the point in my introducing myself to them? There was a clatter on the floor behind me. I turned round and found Haastrup shuffling through some books from one of my shelves. He held one before him, and nodded with a cynical smile on his face as he read the title on the cover.

'What exactly do you do now, Mr Dukumo?' asked Osawe.

'I'm a freelance journalist,' I said. 'Everybody knows that.'

'I know. I've often read your pieces. But is that all you do?'

'I make enough doing that to survive,' I said.

'I can see that,' he said. 'And of course things must be considerably easier for you since your wife left you.'

I resented that comment on my private life. I gave him a stony look, and might have said something uncomplimentary, but my attention was diverted by the sound of something smashing on the floor. Haastrup had upset a porcelain bowl of plastic fruits and flowers resting on another shelf, and the bowl had broken to pieces. Without the least effort at an apology, the son of a bitch was sticking his greasy-looking fingers between the spines of books and inspecting their titles. I made to get up and walk to him, but Osawe held me down with a hand and raised brows.

'If I were you,' he whispered, 'I'd ignore him and concentrate on our discussion. Even I have trouble restraining him.'

'But he's destroying my things,' I objected.

'Never mind. It will all be over soon.'

I turned my attention from the brute behind me and faced Osawe once again.

'What papers do you write for, Mr Dukumo?'

After a brief silence I said, 'Mostly the *Standard*. Sometimes the *Beacon*, *Nigeria Today* and the *Republic*.'

'Is that all?'

I knew where he was leading, and I didn't want to be accused of hiding anything. So I told him, 'A few foreign papers, and magazines.'

'Which ones?' he asked.

'Some of my stuff has appeared in *Drum*, *Ebony*, the *New African* and others.'

'Are they commissioned, or do you send them without being asked?'

'Both,' I answered.

He opened his briefcase and leafed through the contents. He brought out a copy of a February issue of *North/South* in which I had done an essay on the ethnic factor in African development, and opened it up on my piece. After browsing for a while he read out:

' "The scourge of ethnicity in contemporary African politics is starkly demonstrated by the fact that Nigeria, the largest nation on the black continent, fought a civil war whose origins may be traced without any question to ethnic rivalry. Our leaders have failed woefully to turn the ethnic plurality within our borders to good account and have indeed often sought to exploit the inherent rivalries for their own selfish interests, with little regard for the overall welfare of the nation. Is it any wonder that the generality of the elite, taking their cue from the leadership, have ..." Mr Dukumo, why have you chosen to expose your country in this way to the ridicule of the outside world? Aren't there any attractive things about this country you could write about in the foreign press?'

I was going to ask him, 'Like what?', when Haastrup ambled up to us with several volumes from my shelves in his hands.

'*An Introduction to Marxist Ideology*,' he read out. '*Revolutionary Politics. Cry Freedom. The Wretched of the Earth. The Thoughts of Mao Tse Tung.* What are you doing with these books?'

'I read them,' I lost no time in telling him, in a tone calculated to suggest to him the difference between us.

Haastrup looked at me, and I could tell by the meanness in his eyes that he wished I'd said something more offensive than I had. At this point I really didn't give a shit. Osawe must have felt the growing menace and decided to regain the right of interrogation.

'Mr Dukumo,' he said, 'are you aware that there are elements in this country today trying to destabilise the government and throw the nation into political chaos?'

'I'm not surprised,' I said. 'That happens all the time.'

The way they looked at each other, these gorillas, they obviously weren't quite prepared for that kind of reply.

'And that these elements get some help from foreigners?'

'Possibly,' I said, looking Osawe straight in the eye.

Haastrup took over, obviously impatient with the pace of his colleague's questioning.

'You attended a meeting of the International Civil Liberties Association in Ivory Coast last year, didn't you?'

'Yes, I did,' I said.

'Who sent you there?' He now seated himself, with magisterial arrogance.

'I was sponsored by the *Third World* magazine to cover the meeting.'

'Why you?'

'I'm freelance, I've done some reports for them before, and it was cheaper for them to sponsor someone right here in West Africa than fly one of their men over from London.'

'So you attended as a hired freelance writer, a paid stooge. And you did more than just reporting on the proceedings, didn't you?'

'I don't think so,' I said. I wondered what he had in mind.

'Read him his statements.'

Osawe fished in his briefcase again and brought out a typed sheet, clean and fresh. He read out something that sounded like an indictment of the Nigerian authorities for holding about 300 men and women across the country in detention without trial, and for the inhumane conditions in which prisoners and detainees were being made to live.

'Well?' Haastrup cut in at the end of it, glaring at me.

'Not me,' I shook my head. 'I never said such things.'

'Are you calling us liars?'

'No, I'm not. But I don't recall using such words during the ICLA conference.'

Can you believe that? The bastards were trying to plant something like sedition on me!

'Well, *what* did you say? What words did you use?'

They had got me there. But, damn it, that conference took place over a year ago. How could I remember anything beyond the report I wrote for *Third World*?

'Whatever I said,' I replied, 'whatever questions I must have asked, were all part of the report I submitted to the magazine and which they published. I can show you the story if you want me to.'

'We have seen the story,' Osawe rejoined. 'Are you saying you left nothing out?'

'Nothing that I can remember,' I said. 'Of course, as a journalist I make a lot of notes, and then tailor the final report to fit the size of text commissioned by the magazine. But nothing that is really significant gets left out of my stories.'

'Significant, eh? Let's see if you didn't consider this one significant. Read him the next statement.'

Osawe read out something else, to the effect that in Nigeria there didn't exist any climate for a truly democratic exchange of ideas, and there could never be one for as long as there was no constitutionally elected government.

'I *never* made such a statement,' I swore. 'I demand to be shown evidence that I said such things.'

Haastrup lost no time in accepting the challenge.

'All right,' he said, rising up. 'Let's go and play him the tapes. Come on.'

'Where are we going?' I asked, looking to Osawe for possible sympathy.

'To the office,' he said with a sigh, rising too.

I didn't like it. I'd been to *that* office before, you remember? If I went there this time, how the hell long would they keep me? I remained uneasily on the edge of my seat while the two men stood, until Haastrup yanked me up by the arm and barked out:

'Stop wasting our time, my friend! You asked for evidence and we want to show it to you.'

I had on just a long-sleeved shirt and trousers. From a nearby chair I grabbed my jacket, slipped a pair of casuals on my feet, and locked the door as we went out to their waiting car. I had no illusions about being released quickly. But little did I suspect what these men had in mind, either. As soon as we entered the Security Division of the Police Headquarters, Haastrup handed me over to an officer with another hard push which forced me to hold on to the doorpost to avoid crashing to the floor.

'Emiko,' he said, 'let him cool off until he is ready to cooperate with us.'

'No problem,' replied the officer. He pulled some keys out of a drawer and said. 'Oga, let's go.'

When I looked behind me, Osawe and Haastrup had gone. The motherfuckers had actually put me in detention – just like that!

As Corporal Emiko Esimaje was opening the iron gate leading into the detention quarters, I saw about three young men emerge suddenly from a side door down the hallway. God, Tonwe – you should have seen their faces and the state they were in! That was the first fright I experienced at the place. You know the lunatics roaming the streets and garbage dumps of Lagos? That's what they looked like – unshaven, unkempt, filthy, tattered, wild, somewhat like scarecrows suddenly come to life. And they were rushing towards us, each struggling to keep the others back! Esimaje had unhooked the massive padlock from the last of three bolts and was about to release the bolt, but I held his hand on a sudden reflex and took cover behind him. He glanced back at me and burst out laughing.

'You'll get used to it,' he said. 'But, oga, if you love yourself, make you jus' give dem anyting wey dem ask you. Na so dem deh do am here – o.'

Instinctively I touched my pockets – jacket, then trousers. To my relief I found there was some money in one of my trouser pockets. Esimaje unbolted the gate and let me through. As soon as I was on the other side he clanged the gate to. By this time the inmates had got to me and started fighting to gain control of me. One of them, Johnny, overpowered the other two and got hold of my jacket. The others refused to yield and tried to pull him off. In the process Johnny ripped my outer breast pocket. That infuriated me and I bawled out with my hands raised:

'Now stop it, and listen to me.' All of a sudden they halted their stampede and stood still. On realising how easily they cowered to authority, I proceeded to take advantage of the situation. 'You don't have to tear my clothes to pieces. Just tell me what you want and I'll give it to you.'

'All righti, oga,' said Johnny, asserting his position as their leader. 'Jus' bring all de money wey dey your pocket, put am for my hand now now. I know say you be gentleman, but jus' *je-je* ...'

'Na you only one wan' take am?' another one of them cut in.

'Sharrap!' Johnny shouted him down. 'Who firs' reach here?'

'Oya now,' said the third, 'you don try. No be you go take the money sef. You just be messenger like we all.'

Johnny ignored him and faced me again, repeating his demand, his eyes blazing red. I did not argue with him or even try to hesitate. You don't push your luck with such people. I put my hand in my pocket and fished out all the money I had in it – fifteen naira in all. After I stopped, Johnny was still staring at me as though he was sure I was holding something back. I proceeded to turn all my pockets – jacket, shirt, trousers – inside out for him to see. Only then was he convinced I was in earnest. He counted the money in his hand and nodded contentedly to himself.

'I know say you be gentleman,' he said. 'Oya, follow me.'

'Johnny, na how much 'e give you now?'

'Sharrap!' Johnny shouted again at the questioner. 'Na your money? Longthroat. Beast of no gender,' Then he motioned to me, 'Come boh make we go meet our chief.'

I don't think you can guess who the "chief" was, nor can you imagine with what mingled feelings of delight and consternation I found myself face to face with Mr Ebika Harrison! He was seated quite comfortably on a sofa of sorts placed against the wall of a fairly large room, which was to be described to me later by Johnny as the chief's "office." He had on a pair of sinister looking sun-glasses, and his beard had grown even bushier. He was flanked by a pair of grim-faced stalwarts who I gathered later were his bodyguards. As I said, I was pretty delighted to see Bickerbug after such a long interval but somewhat baffled by the atmosphere surrounding him. Still, I couldn't help being familiar.

'Ebika,' I greeted, moving towards him with a cautious grin on my face and a half-outstretched right hand.

As I got closer to him, the two stalwarts stepped forward, almost in unison, and would have laid their paws on me had Bickerbug not stopped them. He didn't show any anxiety, nor did he move an inch. He simply said something I couldn't quite make out. The words were barely audible, but they definitely carried a lot of weight with those gorillas, for they stood back from me instantly and resumed their standing positions beside Bickerbug. I swear, those faces could never have smiled in all their lives, nor the hair surrounding them ever felt the barber's blade. I must confess I was relieved that Bickerbug came to my rescue. I thought I'd been manhandled enough for one day. And the way those fellows looked, if they ever had me alone at their mercy, what I experienced at the hands of Dayo Haastrup would have seemed in comparison little more than a tender pat on the shoulder.

'You are welcome, Piriye,' said Bickerbug with an impassive face and tone. 'Make yourself at home. You'll get to know everyone here soon enough.'

I still didn't have the courage to take up the offer of a seat. He turned his face to one side, and this was obviously a familiar signal, for at once Johnny ran up to him and reported, 'Fifteen naira, chief.'

Bickerbug nodded slowly at the information, then said, 'Take five naira to the treasurer, and give back the remaining ten to the gentleman.'

'Sir?' Johnny asked, with a look of disbelief.

Rather than answer the question, Bickerbug turned his face slowly but with unmistakeable menace towards Johnny. The message was not lost on Johnny, who quickly blurted out, 'All right, sir.' He counted out ten naira of the money and handed it to me, and took the rest away. By this time, of course, the rest of the team that received me at the gate had disappeared – in fact the stampede towards me then had been only a competition to see who would have the privilege of taking my money to Bickerbug. So, my dear Tonwe, in this way I was introduced to the court of King Ebika the Great! The royalty I'm talking about was no laughing matter, as I'll tell you shortly.

After Johnny had disappeared Bickerbug sighed and allowed a mild grin on his face.

'Sit down, Piriye,' he said. 'Relax.'

I lowered myself and sat on the bare but clean-swept floor. I looked around to familiarise myself with the scene. There was nothing at all in the room. It had only a door and a window without frame or flap, and these were its only sources of light in the daytime. Later, in the hours of darkness, it benefitted only slightly from the light of a lone electric bulb hanging down from the ceiling of the hallway with a frayed and twisted wire-flex. As I said, Bickerbug's "office" – for so it was referred to during the two days I spent in detention – was fairly large, and besides the two silent guards flanking him I felt no longer threatened by anything once I was asked by the chief himself to feel at home.

'You can't imagine how long I've been trying to see you,' I said to Bickerbug. 'A few days ago I actually got close to this place.'

'I know,' he said. 'You came asking for me at the Security offices, and they kept you under interrogation for a few hours.'

My mouth fell open as I stared at him in wonder.

'Don't worry yourself,' he said, with another mild grin. 'You'll be amazed how much I know about what goes on outside these walls. You may even be amazed who my sources are. But that's not important. I knew it was only a matter of time before you came here yourself.'

I went on to tell Bickerbug how I finally got here and the entire experience I had with Osawe and Haastrup. All through my narration he was simply smiling, sometimes chuckling and nodding sagely. Indeed at one point he almost repeated with me, word for word, the statements I was reporting my captors to have made in the course of incriminating me.

'Idiots,' he said. 'The trouble with these Security boys is that they lack imagination. Using the same style every time. It was the same boys who brought me here. Just like they did with you, they read me statements I was supposed to have made sometime in Campos Square. Although I never said those things, I didn't deny them. I simply asked them how they came to be so accurate. *Then* they told me to come along to their office so they could play me the tapes. Well, I refused to go with them, and we had an argument and a little scuffle, and they overpowered me, and took me away.' He sighed and shook his head with vehemence, as if trying to dispose of an unnecessary nuisance. 'Well,' he said, throwing out his arms and affecting a smile, 'welcome, my friend, to the world of legalised blackmail. I hope you're learning.'

'I am,' I said. 'I am.'

It was getting a little dark. I looked up at the bodyguards, and found them watching me with not menace so much now as a stern sense of duty. Suddenly I heard an alarm ring, and I looked at Bickerbug.

'Inspection time,' he said, rising briskly from his seat.

There was a bustle along the hallway, as several inmates hurried along in the direction of the gate. I got in on the act at once, and found myself standing in the second of two lines facing the gate. Bickerbug brought up the rear, flanked by his bodyguards and walking at a leisurely pace. He finally took his place in the same line I was in, still flanked by his men. Four armed guards surveyed our lines, counting our number with their nodding heads. Shortly after, my old friend appeared on the other side of the gate – the Northern officer who had grilled me in the Security waiting room the week before. DSP Adamu Yelwa, I was to be told later by Bickerbug. He stood outside the gate with a stern face, his legs at ease and his hands clasped behind his back.

'All present and correct, sir,' one of the armed guards shouted.

'All right,' said Yelwa. 'Dismissed.'

The armed guards stood aside, and we all walked back to our quarters. On the way Bickerbug called Johnny.

'Make sure my friend is comfortable in your cell. I don't want to hear that anyone bothered him.'

'All right, sir,' said Johnny briskly.

Half an hour or so later we were served supper – if you can give that name to the stuff that was given to us. God, Tonwe – you never saw such garbage, nor the sort of plates that it was served on! We had to line up by a tiny window – hardly big enough to allow a human head through it – in the wall separating us from an adjoining kitchen, while a cook or steward slapped the slosh into our greasy plates – some wretched brew midway between soup and porridge – and threw a rugged slab of stone-hard bread into it.

I can't quite say which offended me more, the food or the plate. After I had collected my ration and taken one look at it, I nearly threw up. I walked over to my cell and just sat on the wooden boards on the floor that were supposed to be my bed, pushing the plate away from me and, my head bent over my hunched knees, wondering how I got myself into that terrible mess. When Johnny came into the room and saw my food lying untouched, he rushed to secure it before anyone else could find it.

'A-ah, oga, you no wan' chop your food?'

I simply shook my head and told him he could have it, and he lost no time in slurping it up, stopping every once in a while to wonder perhaps at the crazy man he was going to room with.

Bickerbug was never on the food line. When he heard what I'd done he sent for me, and when I got to his room – again a larger cell which he shared with his bodyguards, though he alone had a bed with a sheet of foam over the boards – he gave me a little lecture.

'Piriye,' he said, 'you've got to be strong. Don't allow these fellows to break you. You are more intelligent than they are. The last thing you want is for them to see they have any kind of advantage over you, because they'll exploit it to the full.'

He gave a sharp whistle, and a young boy – who couldn't be more than eighteen – ran into the room.

'Atila, tell the cook I want the special – now,' Bickerbug ordered.

Atila ran off, and in no time at all I was sitting down to a clean dish full of stewed beans and two large slices of soft bread with a bottle of Coke to wash it all down. This was a welcome treat, for all I'd had for lunch that day was a meat pie and a glass of water.

'That was just to welcome you to the place,' Bickerbug said, after Atila had returned the empty plate and bottle to the kitchen. 'From now on you really have to eat what everyone else eats. I eat it myself. You've got to survive, you've got to live and carry on the struggle.'

'Well, thanks man,' I said, rising up to go. I was tired.

'Relax,' he said. 'We still have a couple of hours before bedtime. Besides, it's nice to see a familiar face in this dump.'

I took up the offer, and rested my head and back fully on the cushioned bed. My eyes happened to meet the mean stare of the two bodyguards. But I'd become used to their menace and could afford to ignore it, especially in the knowledge that, now they knew I enjoyed a special relationship with their boss, they dared not lay a finger on me. I must have dozed for something like an hour, then I was awoken by the sound of gunshots and a loud scream. I raised by head from the bed and stared at Bickerbug with some alarm. He merely shook his head and hissed, saying, 'Bloody fool. He wouldn't listen.'

'What's that?' I asked.

'Some fool who's been trying to escape for some time. I tried talking to him, but there must be a devil driving him. Now they've got him.'

There was a stampede towards the wall behind, and though Bickerbug didn't immediately get up I took off to see what was happening. The light was faint in that part of the premises. I pried my way through the ring of inmates, some of them sobbing and others merely speechless with shock. Inside the circle two of the armed security guards were standing over a groaning man. I looked closely and saw a pool of blood on the floor where he had collapsed. The bullets had ripped his right thigh to shreds and his abdomen was open, allowing his intestines to spill out and hang loosely over his shredded shorts. Presently a third guard shouted, 'Everybody stand aside!' and cocked his rifle.

We all moved aside. The two soldiers who had gunned the inmate down now grabbed him, each by the arm, and proceeded to drag him along the ground in the direction of the main building. Just then Bickerbug appeared with his bodyguards and signalled the two armed guards to stop. They did, and he called two other inmates – Sikki and Adisa – to carry the wounded man more carefully. By now he could barely breathe. Bickerbug asked them to take him to a medical bay on the north side of the building, from where he was to be taken later to a hospital. The guards didn't contest Bickerbug's instructions. He offered a few words of comfort, though without emotion, to the wounded man, and walked as majestically back as he had come, flanked by his two strongmen. The rest of us walked slowly back to our cells, in an atmosphere that was both funereal and tense and with hardly a word said by anyone. I didn't see how the poor man could survive.

I had neither the heart nor the presence of mind to go back to Bickerbug's room. In any case, not long after this the alarm sounded three times, which I was told was the signal for all lights – except the ones around the walls and the gate – to be switched off and all inmates to retire for the night.

I could hardly sleep that first night. One reason was the stench of the toilet pail in our cell. By the way, I should have told you earlier about this method of sanitation. When I got to my cell after the inspection earlier in the evening, the pail was the first thing that disheartened me. Having to sleep on the floor with hard boards was no great surprise – I never thought detention was going to be a bed of roses anyway. But to have a pail of shit right there where you would sleep! It wasn't a big pail, and it was supposed to serve the four of us in that cell – Johnny, myself, and two other inmates one of whom, an older man, was called Baba and the other nicknamed Captain Blood (I never asked his real name). Besides the size of it, this pail was torn for about half its depth from the top, so that there was a constant overspill of human waste on the floor by the corner of the wall where it stood.

There was a lid for it, and I asked Johnny why the lid couldn't be placed on the pail. But he told me the other inmates had insisted I should be introduced to the realities of life in detention and that the pail should be left open for my first night there.

If I had any illusions about resisting that order, I was restrained by anxiety over my cellmates. Johnny was okay – for one thing, Bickerbug had put him in a good disposition towards me. Captain Blood tried to frighten me when Johnny brought me to the room and introduced me to him and Baba. No doubt what irked them was the special treatment that Bickerbug had ordered for me. When I held out my hand to shake Captain Blood's, he slapped it away so hard that I had to clench it to relieve the pain.

'Oppressor, shit!' he spat. 'Keep away from me if you want to stay alive.'

I smiled faintly and stared at him, and for a while we eyed each other. I heard later that he had a reputation for violence. But after the staring match I decided he wasn't such a serious threat so long as I kept my distance from him. I felt too that I'd established a certain respect for myself with the way I looked straight into his eyes and stood up to him. Still, I didn't think I should take any chances with him, that first night at least.

My real worry was Baba. When Johnny introduced me, he said nothing and didn't even look in my direction. He was sitting on the floor and rubbing his hairless head with his hand. Every once in a while he would walk round the room, and the first time I had a view of his eyes I noticed they were bloodshot. In the course of his walk he would stop by a wall, rap on it with his knuckles and listen closely, as if trying to determine the density or hollowness of the concrete. I couldn't figure out the man, so that even when he lay down at night in a position of sleep I couldn't be sure he wasn't awake and up to something, especially since I never heard him snore.

With the shit pail and Baba, it wasn't easy for me to sleep that first night. I folded my jacket to make a sort of pillow for my head, and just stared at the ceiling for a long time. My mind was really on Bickerbug. I wanted dearly to talk with him and find out how much had happened since I'd last seen him. For the moment, though, I was somehow intrigued at how he had come to occupy such a commanding position among the inmates of the detention camp. I wasn't sure I'd ever find the opportunity to hold a private discussion with him for any reasonable length of time, but I was determined to get closer to him and at least ask him a few questions. I had some luck that first night as I lay on my boards staring at the dark ceiling and rolling a myriad things in my mind. It was Johnny who broke my concentration – and I'd thought he'd dozed off!

'Oga,' he said, 'make you no vex – o. But i' be like say our chief like you well well.'

'Why?' I asked. 'Wetin make you think say 'e like me?'

'I never see am treat anybody like this before. A-ah, na wa-o! We deh fear am well well for here – o.'

'Na so I see am,' I concurred. 'The man na tough man.'

This gave me the leeway I needed. I went on to ask Johnny how Bickerbug came to be boss among the inmates, when there were older men than he around – older both in age and in length of stay in detention. Johnny then gave me an account which confirmed the toughness of character which I'd always seen in our man.

Bickerbug had been brought in there about the third week of last month. At that time, there was another boss – by the way, I came to understand that in a place like that there is always a top dog, some inmate towards whom all the others gravitated, either because he inspired the most fear or because he had the personality or clout to secure for the community of inmates a reasonable deal from the authorities. Now, the man who was chief when Bickerbug came into detention was an elderly man of about sixty – a bulky, grim and macho-looking fellow who spoke with considerable fierceness, but who proved woefully incapable of saving the inmates from the brutality of the guards.

It took Bickerbug only about one week of being there to take over control. He had been eyeing up the old man and didn't think he deserved the position he was enjoying. His chance came one day when one of the inmates was so fiercely whipped by the guards that he could hardly rise from the ground – simply because he had been found loitering around the walls of the premises when everyone else was busily engaged, and the guards had thought he was trying to explore an escape route. The guards turned out to be right: it was the same young man that finally got shot on that evening of my first day. But on that earlier occasion the guards had acted only on a vague suspicion and had beaten the young man to a pulp.

Bickerbug had been infuriated by the event, especially because the old chief – by name Okpaleke, Johnny told me – never said a word either during or after the beating of the young man. While everyone else went back to his business, it was Bickerbug alone that helped the man back to his cell, cleaned his body of blood, and solicited medical attention for him.

Later in the evening he took a step he had been contemplating for some time and for which he came fully prepared. Okpaleke had been sitting in the "office" – now Bickerbug's – surrounded by a few close colleagues or sycophants, when Bickerbug walked up to him and asked him to get out of the place. The old man didn't think anyone would have the nerve to talk to him like that, and asked Bickerbug to repeat what he had said. Bickerbug did, adding that Okpaleke was a disgrace to the human race and in any case was too old to face up to the responsibilities which his position had placed on him. The old man made a bold effort to assert himself and, springing up with amazing agility, landed a resounding slap on Bickerbug's face. Bickerbug's response was just as swift – his slap sent the old man reeling backwards several steps. Okpaleke dashed forward to grab Bickerbug by the neck. But the younger man was too quick for him. Out of nowhere Bickerbug flashed out a flick-knife and held it to Okpaleke's throat. The latter was transfixed on the spot.

Everyone else there rose quietly and left the room. Poor old Okpaleke finally followed suit, his eyes carefully watching, now the blade, now his subduer's face, as he backed slowly out of the room.

Regulations at the camp forbid anyone being in possession of any kind of weapon. But no one had the courage to report Bickerbug to the authorities. In fact he had come to command among everyone there so much respect and awe that it was doubtful that the guards, who regularly carry out an inspection of the cells and premises and have the right to frisk anyone they suspect of hiding anything illegal, would have dared to disarm or even challenge him if they'd got wind of his possession of the weapon.

I asked Johnny how Bickerbug came to have those two stalwarts as bodyguards.

'A-ah, na because 'e be our chief now. Chief no deh get guard?'

'Okpaleke been get guard the time when 'e be una chief?'

'No,' he said. 'Sometime 'e no get sense for take ask. But this our new chief – ha ha! Boh, man pass man!'

I then asked Johnny if Okpaleke was still in the detention camp. He burst out laughing and said:

'Na, 'im dey near you so.'

I quickly raised my head and looked at the old man lying a short distance away. The news nearly scared the shit out of me.

'This Baba?' I whispered, lest he might hear me.

Still laughing, Johnny nodded yes. I adjusted my lying posture so as to face the old man, in case he tried anything funny during the night.

'No worry yourself,' Johnny said, his laughter subsiding now. ' 'E know say you be our chief friend. If 'e touch you, chief go show am pepper.'

You can now understand why, apart from the shit pail, it took me a long while to get to sleep that night. But two further shocks were waiting for me when I woke up the next morning. The alarm was sounded early, about 6 o'clock I think. As I opened my eyes, whom did I see but the old man himself, standing over me and staring straight down at me, his red eyes glaring, his legs astride me and his arms akimbo! I recalled I had in fact seen him in my dream, and seeing him there was to me a sort of lingering nightmare. I raised myself quickly and backed away just as fast. Johnny and Captain Blood had observed the drama from their positions on the floor, and burst out laughing. To my relief the old man simply walked away – he didn't laugh or say anything, he just walked out of the room.

The other shock was my money – the ten naira Johnny had given me back the evening before. You must understand that everyone in the detention camp wears the clothes he came there with. They are not prisoners in the regular sense, who have to remove their clothes and put on the prison uniform. They are simply being held there as security risks who have to be kept under some surveillance until they are cleared – although many stay there for an incredibly long time and may just as well have been convicted prisoners serving long terms, only because the authorities have forgotten about their cases or lost their files.

So in the camp you keep your own clothes on. On my first night, as I've already told you, I folded my jacket to cushion my head against the boards we slept on. The next morning when I got up, I put the jacket back on. When I felt for the ten naira in it, the money was gone! I couldn't decide whom to suspect. Was it the old man – could he have stolen my money? Or was it Captain Blood, playing the tough guy with me when he was no more than a petty thief? Or was it Johnny, who after all had spent more waking time with me the previous night and so knew exactly when to sneak up on me – had Bickerbug delivered me into the hands of a rogue? Anyway, I thought there was no point bothering myself over the matter. I might get myself into more trouble trying to find out who stole it – besides, what would I buy with it there?

Tonwe, I won't bore you any more with the routine of life in detention: how the shit pails are emptied every morning, how you can only have a bath once a week and so many other sordid details. I'll simply mention a couple of things about Bickerbug which gave me an insight into this remarkable person I have become inextricably bound to. First, it's amazing how much he had achieved for the camp within only three weeks of being there. As I learnt especially from Johnny, it was Bickerbug who introduced the few sporting facilities available at the camp – soccer, volleyball and draughts. Before he came it was all physical work alternating with long hours of boredom. Bickerbug argued vehemently against the authorities' claims about limited financial resources, trying to establish the point that inmates had suffered enough by being isolated from normal society and deserved to stay alive and healthy. He even threatened to lead a general riot to press their claims, even if it cost all of them their lives. The authorities knew he meant what he said, and had to meet his demands. I participated in one of those soccer games on my second day in the camp: to think that I had to be in confinement to find time for real relaxation!

These games lasted mostly from 4 to 5.30 p.m. – a schedule I heard was also introduced by Bickerbug. Along with the games came another one of the amazing innovations of Bickerbug, something he called the "open forum". This took place in his "office", and lasted between thirty minutes and one hour. These rap sessions took the form of open discussions on a whole variety of subjects involving the life of the citizen. I was told about earlier discussions held on subjects like the system of taxation in the country – was it equitable, was it efficient? Traditional rulership – what use was it in our present circumstances? Women in society – were their rights sufficiently guaranteed, should they enjoy greater prominence in social and political life than they do at the moment? Even subjects like marital life came under analysis – was polygamy sensible? Should husband and wife operate joint bank accounts?

The session I attended after our soccer match was a straight political one. With the recent lifting of the ban on political activities, the inmates apparently didn't want to be left out of the on-going debates on the prospects of the Second Republic: was it right for the government to allow the formation of as many parties as possible? Shouldn't we be content with a two party system? Was the constitution prepared by the Constitution Drafting Committee (CDC) and approved by the Supreme Military Council likely to bring any stability to our political life? Did it not leave too much power at the centre and too little in the states? Wasn't there need for yet more states in the country so as to ensure an even spread of development across the country? What should be done to discourage future coups?

I was impressed at how lively this particular discussion was. I happened to be sitting close to a lawyer from southern Zaria, a certain Barrister Yohanna Idriss, who was being detained on charges of holding political meetings before the ban on politics was lifted, and he told me that earlier sessions had been just as lively. It was to the credit of Bickerbug's organisational power that the whole affair was conducted in as orderly a manner as possible, given the variety of social classes from which the inmates came. I have already said that there were quite a number of inmates older than Bickerbug in both age and length of residence at the camp. There were also quite a few of higher social standing. He has only a first degree in English and until recently was just a secondary school teacher. But in that camp you have figures like a university professor, two company directors, a permanent secretary from one of the states, a former diplomat, Barrister Idriss whom I just mentioned, a high-ranking military officer (Colonel Olanipekun Ikotun) from the National Youth Service Corps, a Director of Customs or something like that – all there on a variety of charges. Now, many of these people were older than Bickerbug, but it is amazing how they submitted themselves to his authority and looked up to his direction of their affairs in the place. His bodyguards stood constantly beside him in the session I witnessed, but he had no reason to use them to impose discipline because even though the arguments sometimes got heated, they were never out of hand.

The session was also thoroughly democratic – everything was conducted in pidgin so that both the educated and the not so educated could deliberate on equal terms and hold a true dialogue. Bickerbug also insisted that, however sensitive anyone felt about a challenge to his position, he had no right to respond with a rude or provocative remark, or to act in such a way as to impose his superiority on a fellow inmate. His point was that if you were so high and mighty you should be somewhere else, not there! In fact at one point during our political session he made one of the bigwigs – the officer from the NYSC, a soldier for that matter! – apologise to some small fellow whose language he had referred to as "street talk".

I could not help being struck by Bickerbug's personality and even moral stature. When he spoke, he enjoyed absolute silence. In his contributions he made a point of impressing it upon everyone that as human beings and citizens of this country they had rights that should be respected – a right to decent (which didn't necessarily mean affluent) living, a right to protection under the law, a right to education, a right to free speech; everyone had an obligation to respect others as he would like to be respected, and the government – whether military or civilian – deserved our respect and support only to the extent that it did not tamper with our inalienable rights. He said many other things, but the bias was always on dignity, respect, freedom, human rights, stuff like that. I had always seen Bickerbug as a revolutionary fighting a Beniotu cause – little was I prepared for the image of a forceful and even charismatic leader of men that I saw in him in that camp. Even the armed guards stood outside the door to listen with rapt interest. I think it was the awe and regard that he inspired in them both at these sessions and in his general bearing at the camp that induced them to do whatever he told them to do – like asking them to turn over the bullet-ridden body of that inmate to bearers he had appointed.

At the end of the session Bickerbug introduced me to the

community of inmates. I was slightly embarrassed, but I was flattered nevertheless to hear quite a number acknowledge that they often read my pieces in the press – some of them even said it was a shame the way I was treated by the *Chronicle*. Bickerbug asked me to lead the discussion the next day on the responsibility of the press, and before I could reply there was a loud ovation in support of the request. One thing struck me, though. On our way out I asked Barrister Idriss if Bickerbug – or "the chief" as we called him – ever introduced in these sessions the cause he was championing (inmates were also encouraged to bring up their own personal problems or cases), and the barrister said no, and no one ever felt inclined or maybe bold enough to ask him. I thought that rather interesting.

Bickerbug and I had our chat on my second night in the camp, after supper. This was the opportunity I'd been waiting for. As before, he had sent for me through the inevitable Johnny. When I got to his room, he was sitting on his bed; with his sunglasses on as usual and with a rather expressionless look on his face. I allowed myself the privilege of flopping down on his cushioned bed without being asked, and he didn't seem to mind. Our familiarity apart, I got the feeling that Bickerbug is the kind of man that respects courage so long as it does not appear to pose any threats to his person or his position. I may be wrong, though. But he let me relax myself as freely as I wanted. Once again I caught the grim watchfulness in the eyes of his two zombies, and ignored them blissfully.

'Ebika,' I was the first to speak, 'you seem very much in control here. What's the secret, man?'

I had a smile on my face as I looked at him, just to be on the safe side.

'Well,' he said, stroking his beard and possibly concealing a return smile, 'like I told you yesterday, you've got to survive. And this is the best way to do it in a place like this. I don't think you know these inmates yet – not really. You've probably seen the lawyers and professors and company directors, and perhaps you think it's a shame that such *decent* men should be brought here. You may be right. But I also like to look at the other side of the coin. Have you had time to ask yourself what *actually* brought them here? I haven't, and I'm not intending to, because they are not my problem. But I'm deeply convinced that most people – probably not everyone, because government has a talent for making stupid mistakes – but most people who find their way into a place like this are capable of doing some amount of harm. If they don't come here with it, then it grows on them.'

'Does that include me?'

'I said *most* people,' he stressed. 'I haven't yet decided what category you belong to.' We smiled briefly at each other. 'So,' he continued, 'I had to strike fast to assert myself above all the others here. You know the old game – the survival of the fittest.'

'I can see that,' I said, less charitably. The son-of-a-bitch apparently hadn't asked himself if *I* was happy staying under his control. If I hadn't been in that camp myself – as much a victim of the system as he was – I might have said he deserved this reward for his megalomania. I think he sensed some uneasiness in the silence after my last statement, and decided to break it.

'Well,' he said, a little more cheerily, 'how's the free world outside? What have you been doing lately?'

'Looking for you,' I said. 'I was also hoping you'd tell me what – ' I stopped to glance queryingly at the bodyguards, then at Bickerbug.

'Forget them,' he said. 'They're absolutely harmless.'

'Are you sure?'

'A hundred per cent,' he nodded. 'I'm not a fool, Piriye. They don't even understand a word we're saying.'

I took him at his word, and relaxed again.

'I thought you might tell me more than I could tell you. Your friends have been most elusive. You remember you and I had an appointment to meet at your place one evening about three weeks ago. Well...' Something kept telling me I couldn't trust those gorillas standing beside Bickerbug. I was going to look at them again before continuing with my statement, but I knew what Bickerbug would say. I simply decided to apply some restraint to what I had to say. 'Well,' I continued, 'when I didn't see you I got worried. I saw Brown on Broad Street and asked about you, but he said he knew nothing of your whereabouts. I was hoping I'd get closer to them and know what was really happening.'

Bickerbug stared at me for a while, and shook his head.

'You amaze me, Piriye,' he said. 'So, after all I already told you about them, you still expect to know from them what's happening? You probably expected them to tell you the truth too?'

'Well,' I said, 'I'm a journalist. I like to talk to as broad a spectrum of people as possible, but that doesn't mean I believe every one of my sources. Besides, I was going to make other moves, but before I knew it I myself got into the security net. By the way, I don't think I told you that my flat –'

'Was broken into, and they took away your papers, and all that.' He laughed at my surprise. 'You should have expected that, coming all the way to Security to ask about a marked man like me. The trouble is, Piriye, you're too conscientious a journalist to appreciate what a vicious system you're fucking around with. But people like me – we have to stay one step ahead of the hunt.'

'Yes, but –'

'Look, Piriye,' he said, 'stop worrying about how I get my information, because I won't tell you. In case you don't know, there's a code of honour even among thieves. You asked what the so-called CCC is up to these days, and I'll tell you a few things. One, they were responsible for my being brought here – told the Security agents I was planning sabotage, though they had no proof of anything. That shouldn't surprise you, anyway. Two – and I bet you don't know this – Tari Strongface has split from the group. The last time you heard about him was in connection with that meeting with the Minister for Petroleum. Well, he's gone – back to Port Harcourt.'

'Are you serious?'

'Do I look like I'm joking?'

'No, sorry,' I said. 'I mean, how come?'

'You see, they're all little people – petty, small-minded, corrupt and all that. As soon as he got his own cut, he split. Frank Segal got him a job as PRO for Freland Oil in their Port Harcourt office.'

'I might have guessed there was something amiss,' I said. 'I'd been told how Brown and Fiabara went over to the *Chronicle* –'

'To deliver a statement, which they later withdrew, and Strongface wasn't one of them?'

I sighed in mingled awe and despair, and sat up. He chuckled at my discomfort.

'Relax, my friend,' he said, 'I've already told you. When you're in my situation, you have to stay one step ahead of the chase. Anyway,' he rubbed his bushy head of hair, 'I never thought Tari would last in this whole doubledeal. You see, of all the rest of them, he was the closest to me. We're actually second cousins, and though he's a cheap, unprincipled little fellow – as corrupt as they come – I think he got a little uneasy with the group when he saw they were going to sell me to the authorities. I suppose – well, blood is thicker than water, so they say. But that doesn't impress me. As far as I'm concerned, the line is drawn between me and the rest of them.'

'What a shame,' I said.

'That's all right. Now, the third thing you should know about the CCC is that the Task Force mentioned in that report from the meeting with the Minister has actually been set up. Again, that's no great surprise. But guess what: Siekpe and Fiabara are members, with some three other people. The leader of the group is a certain Dr Ojoru Gbekele, an Igala man from that Ministry. Then there's a soldier – a major, I think, but I can't remember his name – and a woman, Dr Alice Chume, who is an Engineering lecturer at the University of Lagos. They are scheduled to ... what's today?'

'November 9th.'

'In roughly five weeks, on December 15th, they will be going off to the Delta on the first leg of their assignment. Gbekele and the lady are experts and should be able to tell the truth. But we know that kind of truth – we've seen it all before.'

I was going to rest my head on the bed again, just to relax and absorb all that information. But the alarm rang three times, signalling bedtime. I quickly rose from the bed and stretched my limbs.

'Well,' I said, 'thanks, Ebika. You've given me a lot to chew over tonight.'

'Oh,' he said, 'don't lose any sleep over anything. It will happen again and again. But I'm through with all that silly talking and make-believe, thinking anybody will listen to me. The mistake they'll make is to let me out of here. Man, if they think I've given them trouble, *they ain't seen nothing yet!*'

This time he volunteered a broader smile than I'd ever seen on his face – I mean, he looked genuinely cheerful. I didn't think any of this was funny.

'We'll talk some more tomorrow,' he said.

'Yes,' I said. 'Goodnight, man.'

'By the way, Piriye,' he said, as I was walking away, 'I'm sorry about your wife. I heard about it.'

I stopped dead in my tracks. I opened my mouth and said 'Who told –', but I didn't bother to finish the question. His face, only faintly visible now in the dim light from the hallway, had resumed its expressionlessness and seemed to indicate he would say no more. Besides, I could guess what he'd say if I'd completed the question – all that shit about staying ahead of the hunt, and how he had his sources, and the rest of it. I thought I'd had enough of that for the time being. So I thanked him for his sympathy, bid him goodnight once more and walked back to my cell.

Bickerbug and I never had another chance to talk. At about 3 o'clock the next day, I was sitting at the back of the building with a group which the professor – I think he teaches English or Drama at Ife – was putting together as a cast for his production of Soyinka's *Madmen and Specialists*. I'd never acted before in my life, and it all looked like a crazy project to me, because he had only one copy of the book, and how were we going to get enough copies to learn our parts? But, I thought, what the hell, anything was better than the drudgery of camp life. I'd never read the book, but from the bits the professor – who seemed to be of the radical type despite the calm bearing I'd observed on him so far – from the bits he read to us, I could see it was a politically provocative play. So some six to eight of us were sitting in a circle with the bearded professor (Ojeifere, or a name very close to that) casting us in our roles, when I heard my name shouted from the direction of the gate. I got up and looked down the hallway, towards the gate.

'Dukumo!' shouted Corporal Esimaje from the gate. 'Come to the gate!'

I excused myself from the group and walked up to the man, wondering what the hell was the matter.

'Bring your things and come with me,' said the man.

A few other inmates had observed the scene, and some had even approached us. I had on my shirt and trousers, having left my jacket in the cell. The thought of freedom hardly crossed my mind at the time. I was wondering if I was going to be subjected to further harassment by the likes of Osawe and Haastrup, or maybe DSP Yelwa, perhaps on the strength of new evidence they had trumped up – or could it be they'd found more incriminating stuff in the papers they'd taken from my place?

Many thoughts like these were running through my mind as I picked up my jacket from the cell and walked back towards the gate. On the way I stopped by the door of Bickerbug's office and looked in. He was there, sitting bolt upright on the sofa with his arms folded across his chest. His formidable sunglasses were once more shielding his eyes, his face was impassive, almost hostile, and as usual the grim-looking bodyguards were on either side of him. I was confused and speechless, and he offered no comfort either by look or by word – he didn't open his mouth or move a muscle. I turned away and continued towards the officer at the gate, when suddenly Captain Blood, who had been standing near a small curious crowd outside a window overlooking the gate, burst out in a song, a recent reggae tune:

Downpressor man, where you gonna run to
Downpressor man, where you gonna run to
Downpressor man, where you gonna run to
All along that day?

I was at first startled by the sound, but having already experienced his harassment I summoned up enough composure to ignore him. Everyone else seemed as confused as I was. As I got to the gate still clutching my jacket in one hand, corporal Esimaje opened up the padlocks and bolts, and clanged them back again after I'd passed through. I looked to him for some kind of clue, but all he said was, 'Come with me please.'

As I walked along the corridors of the Security office I suddenly recovered some sense of who I really was and became thoroughly embarrassed. There I was – filthy, unwashed, probably smelling like a two-day-old pile of shit, and in this condition I was being taken to meet somebody?

'Where are we going?' I asked Esimaje.

He didn't say a word. He just led me on until we stood outside the door of DSP Yelwa's office. He knocked, then opened the door and ushered me in.

'Come in, Mr Dukumo,' said Yelwa.

I walked in, and he stood up and introduced me to a baldish

man in a dark grey suit and tie, who looked vaguely familiar.

'Meet Justice Ekundayo Benson,' said Yelwa.

We shook hands – reluctantly on my part – and said all the usual stuff, hello, pleased to meet you, etc. We all sat down. I remembered him as a judge of the Supreme Court or some high court, whom I'd met sometime on one of my old beats at the *Chronicle*.

'Rough time, eh?' he joked.

I didn't. 'Could be worse,' I replied.

'Well,' he said. 'You can relax now. It's all over. Why don't you put on your coat?'

Then it dawned on me I hadn't done that. I thanked him for reminding me, and put the jacket on.

'Mr Dukumo,' said Yelwa, 'we are going to let you go. But we must come to an understanding.'

I gulped, and stared at him. He had on one of those cosmetic smiles he'd treated me to the first time I encountered him in the Security waiting room. I waited for him to tell me what the understanding was supposed to be.

'There has been some misunderstanding between us,' he said, 'but it has all been cleared up, and we are quite satisfied that you are not a threat to security. But,' and his face dimmed a little, 'your friend Mr Harrison – we have strong feelings about him. And here you have to help us.'

I kept on staring at him, and he took a little time before making his next point.

'We have not found anything concrete against him, but in the three weeks he has been here we have become more and more convinced that he is planning some evil. We could keep him here for good, but it is likely that he is in collusion with some other dangerous elements in this society whom we are anxious to round up along with him. All we want is for you to help us keep an eye on him.'

I couldn't believe this. My mouth fell slightly open, and I looked from him to the judge, who simply returned me a faint smile. He, too, unless he was one of those half-assed judicial types we have in so many courts here in Lagos, must have felt a little scandalised and was simply trying to put on a happy face.

'Do you release what you're asking me to do?' I asked Yelwa.

'I know how you probably feel,' he said. 'But in matters having to do with the security of the nation we should be prepared to compromise our personal sentiments to some extent. Now, we have checked you out thoroughly and we know you are not a violent man and that you are naturally averse to violence. All I am asking is for you to cooperate with us in protecting this country from a danger that seems to be facing it.'

'Yes, but you're asking me to betray my friend,' I told Yelwa. 'I am a journalist. Do you want to make me a dirty spy?'

He had been looking down at his interlocked fingers as I said those words. Now he looked across to Judge Benson.

'It's not quite like that, Mr Dukumo,' said an uneasy Benson. 'I think what the officer is trying to say is that you should cooperate with the government to avert a danger whose proportions we can't quite estimate.'

Yelwa sighed, obviously becoming impatient.

'Let me put it another way, Mr Dukumo,' he said. ' "The mistake they will make will be to let me out of this place" – was that not what your friend told you last night? "If they think I have given them trouble, they haven't seen nothing yet" – was that not what he said? You know, Mr Dukumo,' he said, his tone now becoming patronisingly mellow, 'I think your friend also told you he had his sources all over the place, inside and outside here, did he not? Well, what he doesn't know is that we have our sources too.'

That did it, Tonwe. That took absolutely every word from my mouth that I was going to say. Believe me, I felt trapped. I was confused. I bent my head down and tried to reconstruct in my mind the whole discussion I had with Bickerbug the previous night – at least as much of it as I could. I replayed in my mind everything I'd said during that discussion. I couldn't think of anything I'd said that would be considered particularly incriminating or conspiratorial – thanks partly to my suspicion of Bickerbug's guards, but mostly to his playing the smart, all-knowing, almighty ass that he was. Honestly, I couldn't help hating him at that moment – smart-ass, thinking he had the whole world under his thumb! I shook my head, thinking what a pity the whole thing was.

'Well?' Yelwa said, recalling me to reality.

I raised my head slowly and asked, 'What exactly do you want me to do?'

'Very simple,' he said.. 'Just keep an eye on him. We'll be in touch with you, and I can assure you nobody will know what is going on between you and us.'

At that point I considered the alternative to consent. Suppose I said no, and they bundled me back to that awful camp? I didn't see what good I'd be to anybody, least of all to myself, rotting away in there. Wasn't it Bickerbug who said one had to survive to carry on the struggle? Well, I was convinced I couldn't survive *there*, no sir! Bickerbug is tougher than I am, I was prepared to concede. On a sudden impulse I told myself I must get out of here. I certainly needed time to think the whole thing over and plan how to meet them halfway while preserving my integrity. Survive, Bickerbug had said, and that's what I decided I was going to do. But not in the way *they* wanted.

'Well, how do you expect me to keep an eye on the man if he's still in detention?' I asked.

'Don't worry about that,' Yelwa said. 'We intend to release him, in due course. We just want to tidy up a few matters. Then we will set him free. Don't worry about that.'

I took a good look at the man. *Now* I could see how much poison was in that smile he occasionally allowed on his face, and I hated him even more than I'd done before. But I tried my best not to show it.

'All right,' I told Yelwa, rather weakly.

'Good,' he beamed, and I saw Judge Benson adjust himself in his seat with a certain comfort. 'By the way, we noticed your Passat has been out of action for quite some time. It has been towed to a VW mechanic, and you'll be getting it back in the next couple of days. You need to be mobile to help us.'

This was too much for me! I was too overcome to say anything, not even to ask him how they got to know about my Passat. I was relieved just to be able to get out of that office with Judge Benson.

Benson spoke a great deal as we drove along in his Mercedes towards the mainland, but he didn't get very much cooperation from me. I remember him saying he was Lati's uncle, and Lati's mother had told him the trouble I might have been in, and how he went around talking to his friends, and all that kind of stuff. I don't recall thanking him for his pains or anything of the sort. I was too busy thinking.

Part of my mind was on Bickerbug. Does he really appreciate the extent of the ring they've thrown around him? I think he's a bright fellow, but I wonder if he's as bright as he seems to feel he is. Or could it be that he actually knows what moves the Security people have been making against him and has made sure he is, in his own words, one step ahead of the chase? I was particularly baffled at how Yelwa could have followed the drift of the discussion I had with Bickerbug the night before. Who had told him? I wondered about the possibility of bugging, and of course I recalled the gadgets I'd seen on the wall of the room where I'd previously been interrogated. Now, it never crossed my mind that such a thing would be inside the camp – it never occurred to me to inspect the walls. Could Bickerbug have been so confident in his control of the scene at the camp that he never suspected something had been planted somewhere around him? Perhaps on the so-called bodyguards! Poor Bickerbug. Poor little self-assured bastard, fucking with an enemy he doesn't really understand!

Then I thought about myself. What kind of person did Yelwa think I was? Was there something about me that led him to believe I would turn against my friend, against a fellow tribesman? Was it not the same Yelwa who only about a month ago had read me an extract from some article of mine which gave him the feeling I was part of a Beniotu nationalistic movement? So what gave him the idea that I would turn informer against the same cause I was espousing? Was there something about me, about my character, that I didn't know about but which Yelwa, in his professional wisdom and in all his experience of several men like me he's had to deal with – was there something about me he could already read though it was not apparent to me?

Perhaps you, Tonwe, might be able to tell me. In all the years you knew me in the *Chronicle*, did I ever strike you as a fickle man, an undependable man who had the makings of an informer?

I tell you, the man really got me wondering about myself. I hated him more and more, and I hated even more the system that breeds the likes of him. Of course, I'd made up my mind he wasn't going to get anything out of me concerning Bickerbug. No way! We'll all play this game, and I'll show them they can't take me for granted so easily. I'm only out here because I can't bear to go back to that shit-hole of a detention camp.

As we got to the National Stadium end of the Eko Bridge,

Judge Benson asked me which way I wanted to go – did I want him to take me to his sister's place (meaning Lati's aunt, who had got him to organise my release), so that at least she would know I was now out of detention? I thanked him, but said I thought I should go home first and clean up. So he took me home to my place at Bode Thomas, and I bade him goodbye. Priboye was the first person to visit, and went out to fetch me some lunch. Then he left me to sleep. Lati woke me up around 6 p.m.

So here I am, Tonwe, investigative journalist turned security agent! If there's one thing my experiences within the last month or so have done for me, it's to give me a better insight into the horror of the system we live in. I haven't told you half of what I saw in that camp, nor even half of what I heard from the couple or so fellows there I spoke with, including Johnny and Yohanna Idriss. I'll do so in another letter. It all constitutes an illuminating complement to our project – at least, I can see sufficient material there for a chapter of our book that could be appropriately titled something like 'The Bowel of the Beast'. Bickerbug and his ill-starred kingdom would surely feature as the centrepiece of such a chapter devoted to the morbid counterplay of powers.

Well, I am alive and well – and *surviving!* At least I have no fear now of being seen in Bickerbug's company – and I have no doubt he'll be at large soon enough. Priboye will be bringing you my bundle of letters next week, which is when he tells me he'll be going your way again. From them you'll get ample information about what's happened between me and my wife – or, rather, ex-wife. It's a matter I'm becoming increasingly happy to have put behind me.

My very best wishes to you, Tonwe, and to Madam and Boboango. Tell him my offer still stands. I expect to hear from you when Priboye returns to Lagos.

Cordially,
Piriye.

Brisibe Compound
Seiama.

13 December, 1977

Dear Piriye:

Your letters kept me quite busy. But they also caused me a great deal of concern, and not a little sadness.

The sadness is that you and Tonye could not find it within yourselves to be reconciled. My wife was particularly upset by the news. She remembered the role she had played when you two were starting your life together. How Tonye constantly came to her, to seek advice over this and that. How she was very happy to play the role of a big sister, or indeed of a mother guiding her daughter in her first steps through married life. I myself am sorry that all my pleas, that you consult our Beniotu elders in Lagos for a settlement, came to nothing.

Well, what can I say? If it was the wish of both of you to end the union before further harm was done, perhaps it was just as well. It may have been God's wish too. I hope, however, that you two never have to go to court over this matter, though she has threatened it. I hope some convenient legal settlement can be worked out, that is, if you have truly decided that it is all over between both of you. Please consider that seriously.

I was also disturbed to find how much trouble you have been through in the course of your investigations. The detention I found most unjustified. I know justice is not a highly favoured concept in our society. Why, you and I are vivid proof of the fact. But to throw a journalist into detention, simply because he was seeking information on a detained man, and to trump up charges against him only to justify arresting him: all this is to me most deplorable. I remember that in my earlier years in the profession, as a roving reporter, I had some unpleasant brushes with the authorities. But this is simply beyond reason.

I am sorry you had to go through all this. Although I feel as determined as you are to see justice done over this whole problem, if you do come to feel that our exchange of investigative reports should be dropped, or at least suspended, please do not hesitate to let me know. At any rate, do be very careful in the new role the NSS has imposed upon you. I trust your judgment, but try to exercise caution, and especially control – of your temper, your movements, everything. Be particularly careful about Mr Ebika Harrison. I see nothing but trouble on his path.

The team from Lagos will be most welcome. But I would not be surprised if Batowei chose not to include me in his plans after all, considering our views on the matter differed so widely. But don't worry. If the mountain does not come to Mohammed, Mohammed will go to the mountain. Justice will be done. God does not sleep.

Let me stop here, until I have more to report. Please take care of yourself. I cannot stress that often enough. My wife is still a little lost for words.

Incidentally, may I say, give my regards to your friend, the young girl from the *Chronicle*, who seems to feature rather frequently in your reports?

Merry Christmas!

Yours sincerely,
Tonwe Brisibe.

Lagos, 26.12.77

My dearest, most amiable, most beloved, most respected, most wonderful, most adorable, Tonwe!

Oh Tonwe, you can't imagine how happy I am to be writing you this letter. Oh, my most cherished friend, you cannot imagine how happy I truly am now! This is the best Christmas I have ever spent in my entire life, and I am sure you will rejoice with me, when you know why.

You asked in your letter about Lati, and I am pleased to say she is very very *very* fine! Yesterday, Christmas day, she gave me the most wonderful news I ever heard in my whole life – she said she was sure she was pregnant! Oh, Tonwe, you have no idea how wonderful that made me feel. It was as if I was born again, as if I was totally recreated, as if I had grown a hundred feet taller, as if – oh, Tonwe, I swear I can't describe the feeling to you well enough. I embraced her, and felt a new kind of warmth radiating from her, a new warmth transmitted from her right into me, changing my whole being. To know that I am not "empty" after all, to be made to feel that I am a man like all other men, not just a walking shaft of hollow bamboo – this is a feeling that is new to me after over a dozen years of marriage to a woman who made me feel so much less than a man. I hugged her, I kissed her. I touched her belly, and it felt warm, warmer than it had ever felt before, and I knew there was a part of me growing in there. I moved back a little and looked at her, looked at her belly, and even though it was not that heavy I was sure it had added a bit of weight. Tonwe, oh ...

I'm sorry if I did not prepare you sufficiently in my previous letters for this sort of news. I was afraid you'd think Lati had something to do with the deterioration of affairs between Tonye and me. Nothing could be further from the truth. In fact, I saw her very rarely after we left the *Chronicle*, mostly because for a long time I didn't want anything to do with that paper or anyone connected with it. It was my investigation of the subject of our project that really took me back there, and I needed contact only with someone like her who had worked closely with me. Really, our affair got serious only about ten weeks ago, and I must confess that her company gave me a much-needed comfort from the frustrations I suffered during my life with Tonye, although it was only within the last week or two that I have encouraged discussion of my marital problems with Lati. I say this so as to put the record straight, and I know you will agree. I'm telling nothing but the truth. But believe me, Tonwe, even if you hadn't asked about Lati in your last letter, I'm too overwhelmed with joy to have omitted to ask you to share some of it with me and to pray for us that we see our hopes materialise.

'Pray!' you will exclaim, to hear me say such a thing. Yes, Tonwe, and I mean it now more than I ever did before. This is one change I'm glad to say Lati has brought to me. She has taught me the true meaning of love, warmth and decency – the true meaning, not the pretence of it. In the first five years of my marriage with Tonye, we went to church every Sunday and pretended to be a decent couple very much in love with one another. People really believed we were very close, and they must have seen us as a model of a family that had persevered to stay together in love under God's protection despite the lack of a child to bless the union. Well, you yourself know what has happened in the last several years. All that pretence couldn't hold, because it was not backed up with warmth and mutual regard.

But Lati cares for me – she *cares* about me, and it has made such a difference. I know I'm a restless person and I'm driven by a little too much zeal. Tonye hardly ever protested, and at first I thought she tolerated it all as my God-given nature. It turned out she was really ignoring me – for all she cared, I could go ahead and become consumed by all the fire and energy in me. The result was that my temper got worse, my language got worse, my outlook on life got worse – everything got worse – as I became more and more frustrated by life with her.

But Lati laughs with me, is happy when anything cheers me, and comforts me when she suspects I'm down – for she keenly watches me even when she doesn't seem to be doing so. And she cares about my life-style. She tells me when I dress out of order. And you know what? She has commented now and then on my language by saying something like, 'Mr P' – by the way, she insists on continuing to call me that – 'Mr P, do you really have to use those four letter words?' Well ... you know how old habits die hard. I really do try to curb them, but whenever my zeal or my temper carries me away and the word forms in my mouth she says, 'Ah-ah-ah-ah!', and presses a finger to my lips with the most cheering smile in the world. I swear, I love that girl! She has brought back good spirits and laughter to my life.

And talk about spirit – has she got it! I thought I had zeal and energy, but I've met someone who's more than my match in more ways than one. Take my release from detention. It was only under persistent pressure from me that she told me how her uncle Judge Benson had brought his weight to bear. The old man had been quite reluctant to get involved – even after his sister (Lati's auntie) had interceded on Lati's request – on the ground that he did not wish to appear to be obstructing the course of justice. But Lati went over to his house and asked him a simple question, polite but determined: 'Uncle, does the course of justice mean more to you than the happiness of your own flesh and blood?' Oh, Lati ...

And her job – when she has to get a story, does she *go* for it! You yourself recall how often we commended her on the thoroughness of her coverage. She presses her sources so hard, and is prepared to go anywhere and at any hour, until she gets what she wants. Of course we could say that being a woman helps to bring down some of those barriers that male reporters frequently encounter. But this is a woman with a difference, one who lets her subjects know, without the least touch of offence, how important it is for them to speak out and how much it is in their interest to do so!

Does she have *spirit!* Sometimes I fear she has too much. You recall my telling you in one of my reports how she wanted to take on Bickerbug on her "beat"? Well ... she simply asked to go with me to the NSS to enquire after him, but I feared for her zeal and thought I should at least keep the risks to myself. Does she have *spirit!* Lati – I could tell you about her for ever ...

Please pray for us, Tonwe. I love what's happening to me right now, and I'll go all the way. I'm fully aware that this may crop up as a weapon in Tonye's case against me – that is, if she finally carries out her threat to file a divorce suit. But I don't care. Who would care, if he had to choose between alimony for a heartless woman and the blessed fruit of true love?

I'm sure the oil pollution team will come to you, if Batowei's word is worth anything. By the way, I've got my car back. Isn't it wonderful, how things seem to be working out for me upon my release from detention? First the car came on the 15th, driven to my place – with my own keys! – by a policeman. And then the great news from Lati. Not that I'm excited at the "favour" the NSS have done me – a Greek gift, if ever there was one – which explains why I didn't write to you about it. But with the car I drove with relative convenience to the Ministry, and Tolu Adeoye confirmed everything Bickerbug had said about the team, including the date of their departure to Bendel State. So they'll see you, or you'll see them. Either way, I'm looking forward to your report on developments there. All is quiet so far on this front, especially since Bickerbug hasn't been released yet. And I know better than to go back to that godforsaken office and ask any more of my brilliant questions.

Meanwhile, Tonwe, please rejoice with us and pray for our good fortune. I realise that this union with a non-Beniotu girl may not go down well with our people, but they should consider too that I am entitled to some happiness in my life. Tell madam to bear with me. Tell her not to take it so hard that Tonye and I have gone our separate ways. Who knows, maybe it was God's wish, and Tonye may well find for herself the sort of joy that has come my way. Tell madam to pray for me too: wouldn't she be happy to hear in a few months' time that I am the father of a bouncing baby boy?

My very very best and most heartfelt wishes to you for a Merry Christmas!

Most warmly,
Piriye.

3
FLOOD

Alas, poor me
Woe, I'm done for!
Alas, poor me
Woe, I'm done for!
Father fenced the fishes into
a lake
But the rain-god has burst
its banks
And all the fishes have
disappeared!

Ijo folk song

Brisibe Compound
Seiama.

12 January, 1978

Dear Piriye:

Priboye brought me your letter about a week ago. Let me start by assuring you that my wife and I are just as happy as you are about your good fortune. God closed one door and opened another for you: who are we to question His wisdom? If happiness comes to you through a woman who happens not to be Beniotu, that must be God's will. My wife and I wish both of you well. We will remember you in our prayers. May the good Lord bless you with the child that seems to be on its way. May this be only the first!

The oil pollution team have been here. They have confirmed my initial misgivings, and they have, much more painfully, shaken some of my deeply held convictions rather rudely.

The team picked by the Federal Government came to Benin on 15 December or thereabouts. After two days of consultations with the state governor and his officials, including Batowei and his men, they moved down to Warri. I was sealing my fishing boat one evening in my backyard, when I was told by my wife that I had a visitor. He had come from Benin with a letter from Batowei. I was being invited to appear at the inaugural meeting of the Task Force on Pollution in the Oil Producing Areas, to be held in the Conference Room at the National Oil Corporation in Warri, at 10.00 a.m. on 4 January. I would be paid all transportation expenses and a sitting allowance on my arrival for the meeting.

I excused myself from the young man and took my wife to the bedroom for a brief consultation on the matter. She thought there was no harm in my going, since I had already committed myself to the problem so far as to meet Batowei about it in Benin. I said it was all right. I returned to the young man to thank him for his pains, offered to entertain him. He regretted he could not wait; the boat he had hired was waiting impatiently for him; his journey was long. I accepted his excuse, saw him off in the boat.

This was on 28 December. The following day I went to Opene to tell him what had happened. I asked if he wished to go to

Warri with me. Not that I expected him to agree: after all, he could not follow the discussion once it got under way. But I wanted him to know I cherished his confidence. He thanked me for asking him, but declined the invitation, saying he did not want to cause me the same trouble he did before the naval commander on our previous visit. I had hoped he would say that. I agreed with him.

On 3 January I went to Warri, spending the night with my daughter Enikeye and her husband. The following day he dropped me at the NOC offices at 9.45 a.m. The meeting did not start until about 11 o'clock; we had to wait for Batowei and the Minister for Petroleum and Power to arrive from Benin.

Let me be brief. The group was larger than I had thought. Besides the Minister and Batowei, there were Commander Adetunji (my old friend); the Offshore Manager from the Port Authority at Warri, Mr Oluremi Ketiku; the Exploration Manager from the NOC, Mr Etim Essien; a young lawyer from Batowei's ministry in Benin, whose name has escaped me; another legal officer from the ministry in Lagos, a lady, whose name I have also forgotten. Then there was the team that Ebika Harrison mentioned: Dr Gbekele, Dr (Mrs) Chume, Major Haroun Ismaila, and your friend Brown Siekpe. Ephraim Fiabara was not there, I don't know why. Also present were the representatives from the foreign oil companies: Frank Segal of Freland Oil was there, accompanied by his Area Manager here, Mr Gbubemi Omatsone; Mr Francisco Cioffi, from the Italian oil firm of Buonarotti; M. Pierre Artaud from the French firm of Soritel, also accompanied by their Area Manager here; two other white men from Atlantic Fuels and Pacific Prospects.

Finally, there were two chairmen of councils, one from the Rivers State, Mr Fineface Owubokiri, and our own local chairman, Chief Bieye Zuokumor. I was not surprised to see Zuokumor there, but I was sorry. He is well known for his corrupt liaisons with the oil prospecting companies. This is why the fishermen have never bothered to report their plight to him and have instead sought my intercession on their behalf. He did not conceal his resentment at seeing me there. When our eyes met as we took our seats, he blinked hard and looked away, his face contorted with hate. But that did not bother me.

The Minister, Dr Adiele, and Batowei were at the high table, flanked by their legal officers and Commander Adetunji. At the table immediately below them were the other members from Lagos. The rest of us were seated at a horseshoe-shaped table which occupied most of the space in that Conference Room. The three secretarial staff occupied a table by the wall, to the right of the high table. Some pressmen were in attendance also: there was a young reporter from the *Chronicle* whom I had not known before.

Dr Adiele opened the meeting with a prepared speech of moderate length. He was a stocky man, dressed in a navy-blue Diala tunic and red cap. His voice was thin. He welcomed all of us on behalf of the Federal Military Government. He gave the background to the formation of the Task Force, and the objectives which it was set up to achieve, namely: to ascertain that the standards of environmental sanitation recognised worldwide in respect of oil exploration were observed by the oil companies operating in this area; to allay the fears of the local population about the health hazards posed by the petroleum and gas derived from the area; to explore the basis for possible compensations to be paid to any persons or groups adversely affected by the explorations; and to ensure that both the Federal and State governments on the one hand and the oil companies on the other assumed a collective responsibility to aid development and social welfare projects undertaken by the communities whose territories were being exploited.

He however reminded everyone that the Federal Government was committed to petroleum as the mainstay of the nation's economy, its chief source of wealth. He appealed for the cooperation of everyone concerned, both the oil companies and the communities in which they worked, towards the achievement of the government's development goals, of which petroleum was seen as the principal key. Finally, he warned troublemakers and saboteurs that the Federal Government would deal ruthlessly with any persons who took the law into their own hands and sought to impede the realisation of the objectives the government had enunciated. He thanked us for answering the nation's call to duty in this regard, wished us God's guidance and a speedy progress in our deliberations.

When he had finished, we gave him a loud round of applause. He introduced the officials sitting with him on the high table, and the others from Lagos on the table below him. He then invited the rest of us to identify ourselves for the certification of the list which the secretariat already had before them. We all rose one by one. We gave our names and the places or organisations that we came from. When it was his turn, Chief Zuokumor took a great deal of time to introduce himself, pronouncing his credentials both as a titled man and as the appointed spokesman of our community; anyone else who claimed to speak for the Beniotu of the Bendel State in that room was an impostor who should be treated with contempt, indeed expelled from our midst. I knew he was referring to me. But I did not show any expression on my face. In fact, he caused some laughter among the assembly with his tediously ostentatious claims to authority. When it was my turn, I simply gave my name and my village.

After the introductions, the Minister called on the representatives of the oil companies. They were to state their case and declare the plans their companies had in respect of the objectives stated by the government. Frank Segal was the first to raise his hand. He was recognised by the Minister. He had a pile of papers in front of him, at which he looked from time to time as he made his presentation. He spoke for a rather long time about Freland's international reputation, both as a long-established company with the most sophisticated exploration and refining technology in the world, and as a committed partner in progress of every community that they had worked in across the globe: from Alaska and Texas to Nigeria and Gabon, and all the way to Southeast Asia. He cited their ratings on the New York Stock Exchange, and the numerous Presidential awards they had received for their excellent record. He mentioned the various African heads of state who had thrown their territories open to his company for exploration before anyone else was given a chance to join in the exercise. He even took time to mention which of these heads of state had been given special receptions at the White House in Washington and been made special citizens of various American cities, purely on the recommendation of Freland Oil who had underwritten the expenses of their trips.

He concluded with their programme in this country: the large number of scholarships they had given to students to study courses of their choice, both here and in the United States; their plans to build paved roads as well as schools, churches and community halls in the places where they worked; their willingness to assist the country in the construction of a fourth refinery in any state of its choice, and in the exploitation of the natural gas being wasted in the widespread flaring across the Delta country; finally he announced his company's determination to tackle the problem of oil pollution with the most advanced and flawless detergent technology available. Once in a while he waved certain papers in which he claimed the plans and statistics were contained. I believe he had no intention of letting anyone have a look at those papers; he simply displayed them for dramatic effect. When he had finished speaking, he was loudly applauded.

The other white men, Artaud, Cioffi and the two others, spoke in a similar vein, making very much the same sort of promises, but with very little of the dramatic style of Frank Segal. I think they must have felt that further displays along the same lines would invite doubts among some of us about the sincerity of their claims. So, mostly, they spoke tamely but with a zeal that appealed to our trust. I did not think they were any more honest than Segal. Their plans were no less grandiose, no less familiar to anyone who has followed their protestations in the press for several years.

The Area Managers did not say anything, of course. They simply aided their bosses by supplying them with the relevant statistics as they spoke. This made me even more certain that Segal was just bluffing when he rattled off statistics with so much ease: if there were so many figures involved in these plans, how come he never expressed any hesitation in his recitation? I was not convinced his figures were real. He knew his act well: he was simply smarter than the other executives, better prepared for the scenario.

When the companies had all spoken, the Minister asked for contributions from the rest of the participants. The press asked a few questions. Mostly, they wanted to confirm the statistics proffered by the oil companies. One or two asked about dates. Only one journalist asked a really sensitive question: that was the boy from the *Chronicle.* He questioned the sincerity of the oil companies' claims about the number of barrels of oil a day lifted from our shores; wanted to know if it was true, as generally alleged, that several oil companies colluded with Nigerian middlemen in bunkering, even though the practice had been pronounced illegal by the government. I was proud of that boy, I must say. The question had been directed at Segal, naturally. He looked uneasy. But before he could answer, the Minister intervened to say that the Government monitored all exportations closely, that no reports had reached them of any more bunkering activities. Cameras were clicking here and there.

After the pressmen had exhausted their questions, and no one else seemed inclined to ask any, I raised my hand. The Minister recognised me. I thanked him. Again, my question was directed at Segal. After all, he made the most claims.

'Mr Segal,' I remember saying, 'please tell me a little more about those detergent chemicals you said your company planned to use for neutralising the oil spillages. Is that the only method you have for tackling the pollution?'

'Well,' he said, shifting in his seat, 'there are quite a few other methods: air flotation, the activated sludge process, oxidation, coagulation, evaporation, incineration, trickling filters, gravity separators, you name it. There's a whole lot of technology used in the treatment of oil waste in the environment and I don't want to go into a discussion on that because it's much too technical. One thing though: we couldn't use some of the methods we apply on the high seas in other parts of the world, like environmental sifting of the slicks, because of the complex nature of the ecology in the creeks here. So we've decided that the best method for the sort of environment you have here is the high-density detergent which can be applied to penetrate as wide an area of the mangrove ecology as possible. It's a method that's been given a high rating by the API, the American Petroleum Institute, and I guarantee you maximum success with it.'

That seemed fair enough. But it did not quite dissolve my worries.

'How often will this detergent be applied?' I asked him.

'What d'you mean?' he asked, squinting at me.

'What I mean is this. These spillages are a frequent occurrence. Nearly every week there is a seismic explosion, followed by a discharge into the waters. Or a leakage from a pipeline or a tanker. Is there a guarantee that the detergent will be applied as often as this happens, to avoid the pollution that might follow?'

'My company,' he said, 'has a well-acknowledged safety record in our daily operations. But I can assure you, as we have assured the Federal Government, that we'll do the best we can any time there's any kind of problem. You have my word on it.'

How could I argue over this? These are assurances we have heard many times before. If I raised the point, we might go into an argument, for which I had no statistics beyond our people's concern. So I dropped that line of questioning for another.

'These chemicals,' I said, 'how safe are they for the environment? I am thinking particularly of the fishes in the water. Because it would be a shame if we got rid of the oil slicks cutting off air from the fishes, only to replace them with chemicals that may kill them just as surely. Or, if they do not die at once, they may carry poisons which will become health hazards for those of us who eat them. So how safe are these chemicals for the fishes and for us?'

'The waters are in constant flow from the creeks to the sea,' he said. 'The chemicals aren't gonna be around long enough to do the kinda damage you're talking about.'

Again, I could have pressed on to say that, if the drainage to the sea was as fast as he implied, we would not be complaining about the lingering pollution. The evidence is there for everyone to see.

'What about our farms?' I asked him. 'The waters that irrigate them have often brought the spillage along with them. The crops hardly grow. Will the detergent chemicals have any effect?'

The power had failed. The air conditioners in the room had gone silent, the windows thrown open. Segal was uneasy. He unbuttoned his collar and slackened his tie.

'Mr Er ...?'

'Brisibe,' I volunteered.

'Right,' he said, jotting it down. 'It seems to me you are blaming us for the fact that water flows to your farms. I think you oughta direct your complaint to God, 'cause he's the one that makes the waters flow.' There was general laughter at the joke. 'Or maybe you should consider the alternative: move your farms a good distance away from the waterfront so you won't have to worry about the spillage.'

'That is the next point I was coming to,' I said. 'Seepage. You should appreciate that we need to have our farms not too far from the shores so as to benefit as much as possible from the irrigation. But however far from the shores we have the farms, we are essentially a riverine people. My village, for instance, is a small island. There is no spot on it too far from the waters to be affected by them. And I am suggesting that the petroleum slicks affect not only our fishes and our farms but even the water in our wells. My wife and I have observed an unnatural taste in our drinking –'

'Excuse me!' It was the voice of Chief Zuokumor, booming from his end of the table, his hand up. 'Excuse me, Honourable Minister!'

He had obviously been tolerating me for some time, but could not take it any longer. He even rose from his seat.

'Do I still have the floor, sir?' I asked our chairman, the Minister. Before he could answer, Chief Zuokumor had gone ahead with his interruption.

'I want to know who gave this man the authority to speak,' he said, adjusting his garb, rapping his knuckles on the table.

'I beg your pardon, Chief,' said the Minister, 'but he asked to speak, and I recognised him.'

'That is not what I am talking about,' the chief returned, shaking his head stubbornly. 'Who is he representing? Who sent him here? Everybody in this room is representing one organisation or community. So who is this man representing? He cannot come here and just waste our time talking nonsense, without any mandate from anybody. If anybody here gave him mandate to come and speak for his people, let that person stand up and say so, and he will have to tell us who gave him the right to impose a representative on our people without first consulting us. Honourable Minister, sir, I think this man should shut up because he does not know what he is talking about.'

'Do I have the floor, Mr Chairman?' I pressed, quietly.

'What floor?' Zuokumor thundered at me. 'To sit there and talk nonsense? Talking about tasting petrol in his water. Nonsense! My house is very close to the river, and my drinking water comes from the ground. But I have never tasted any petrol in my water. My Honourable Minister, I say he should shut up at once and let us discuss another matter. He has no mandate.'

At the end of his fulminations, Zuokumor sat down heavily, almost ceremoniously. I happened to look in the direction of Frank Segal, who was sitting near the Chief. He was smiling and nodding with gratification, even patting Zuokumor on the shoulder.

My hand was up.

'As I was saying, sir ...'

'Just a minute, Mr Brisibe,' said the Minister. He had turned to Batowei on his right. They were speaking in whispers. Then Commander Adetunji leaned over to the Minister and whispered something also. The Minister shook his head, adjusted his spectacles, cleared his throat.

'Er,' he began, 'I don't think there is much point in continuing this debate. I'm sorry, Mr Brisibe, but let us suspend the matter for the meantime. Since this is the inaugural meeting of our Task Force, we do not need to have a full or prolonged session. The oil companies, who are the target of our worries, have made their points, and we have noted them. I think the best we can do now is to set up a small working committee to study the depositions of the companies and monitor the situation concerning oil exploration and the environmental problems posed by it. If this is all right with you, I would like to have volunteers for a committee of three or four among you to undertake such a responsibility.'

As soon as he had said that, Zuokumor raised his hand. There were several other hands up, including mine. Before the Minister could recognise anyone, Commander Adetunji leaned again towards him and engaged him in further whispered conversation. At this point, I looked at Batowei. But he studiously avoided my eyes, pretending to take an interest in the exertions of the pressmen and the secretarial staff. Then he lowered his forehead on his interlocked fingers, looking down on the table. I was shocked by the humiliation I had suffered. But I tried to be calm.

The Minister finished his deliberations with the navy Commander, then straightened up and faced the rest of us.

'Ladies and gentlemen, I think we will have to avoid taking volunteers, so as to prevent a possible conflict. I shall seek proper consultation on the matter and come up with a list. The secretariat already have your names and addresses. As soon as a decision is taken, we shall get in touch with those concerned. Meanwhile ...'

'When are we going to hold the next meeting of the Task Force?' It was Mr Owubokiri, the man from Rivers State.

'That's just what I was going to talk about. I am afraid I will not be with you for your next few meetings. From now on, the meetings will be chaired by the Bendel State Commissioner for

Health and Environmental Affairs, Mr Batowei,' he said, turning to his right. 'I leave the matter for the meantime in his able hands. He in fact is the man from whom the letters appointing members of the monitoring committee will come. I also regret that I cannot stay to be with you at lunch. I have to report for a meeting with the Head of State tomorrow morning, so I must hurry back to Lagos today. Let me thank all of you once again for agreeing to serve on the Task Force. The responsibility which you have undertaken is an important one, indeed a crucial one. The country is looking up to you with high expectations. Thank you, and God bless.'

We applauded again. The high table rose. The rest of us rose, allowing the Minister and the other officials to file out. As he passed, the Minister smiled, nodded and waved, even shook a few hands. Cameras were again clicking. Pressmen ran after him to ask questions. I must say, I felt no resentment towards the man. He did what he came to do. No matter what interests he was there to serve, he at least did not treat us with disrespect.

But you can imagine how I felt about Batowei. When he was filing out with the Minister and the others, he took care not to look in my direction, although I followed him with my eyes. By the time the rest of us emerged from the Conference Room, he and the Minister were disappearing in a convoy of official vehicles heading towards Benin. What could I do? Frank Segal too. For a while he had cornered the navy commander into a chat that looked suspiciously conspiratorial. Mr Ketiku of the Port Authority was with them, listening intently, nodding occasionally. After a while Segal and the navy commander drove off in their different cars, but in the same direction. Ketiku rejoined us.

Let me not bother you with the rest of the afternoon. We were served lunch, the rest of us, that is, at the Senior Staff Restaurant of the NOC. I chatted heartily with my colleagues as if nothing had happened. Chief Zuokumor ignored me completely. I did not bother him either. Later, after lunch, I was invited to the Conference Room by a young man. There I was paid my allowances, along with a few other people.

I had contemplated saying no to the money. But I realised that doing that would not do my reputation any good. I had been insulted and humiliated before the whole gathering. The man who had invited me to join the Task Force could not summon the courage to own up, even in a whisper to the Minister, that I had my mandate from him. He had simply allowed the navy commander, who had obviously made up his mind about me from the moment he saw me in the room, to treat me like dirt, to override my opinion.

Still, I was prepared to give Batowei the benefit of the doubt. After all, what would he gain from standing up for me and getting into an awkward confrontation with Chief Zuokumor, who showed no regard for decorum? I decided I could excuse Batowei for the meantime. But what did Commander Adetunji have to be afraid of? What did he have to hide? What was he up to? What was Frank Segal doing holding a tete-à-tete with the navy commander, and getting so chummy with Chief Zuokumor in the presence of everyone? I am not so naive as not to know how sneakily these foreign entrepreneurs dance around our Nigerian officials. I am not even so innocent as to marvel at the immodesty of Americans like Frank Segal who think they have got the world under their thumbs. But I resent the ease with which everyone there seemed – *seemed*, mark you, for I have no way of knowing how the rest of my group felt – I resent the way in which I was made to stand so hopelessly alone by the rest of the Task Force. What, in the name of God, is everybody in this country up to? Is there no honour left in us?

I have been back from Warri about a week. I am still waiting to know what Batowei has in store for me: whether they still have any use for me, or whether I am going to be dropped as a bad case. After all, I did indicate my interest in serving on the monitoring committee, although it has been decided (thanks to Commander Adetunji, I must add) to use a different method of selection. I shall wait a little longer to see what happens next. I am a retired man. I can use all the time I can get. I would be glad to be at peace with myself. But what peace can I expect if I watch my homestead go up in flames before me and not do a thing about it? May God help us.

We are well, and wish you continued good fortune. Please try to be careful, though I feel increasingly less qualified to offer such advice. You will hear from me in due course.

Yours sincerely,
Tonwe Brisibe.

Lagos, 28.1.78

Dear Tonwe,

News of your meeting at Warri came out in the papers several days before I got your letter from Priboye – specifically, I read about it in the *Chronicle* on the 7th. The reporter is a new hand there, Lati tells me, just recently graduated from Ahmadu Bello (Political Science). He cannot have known who you were, or he would have paid his respects, I think. Not a bad report. Naturally, your letter was much fuller and more illuminating – his was probably given a fine "shave", in the good old way.

I have also heard some more about the Task Force from Mr You-Know-Who. By the way, you no longer need to warn me to be careful. If there's anything that life in the last few weeks – detention by the NSS, forced appointment as a secret agent, life with Lati, in its own sweet way – if there's anything these various experiences have taught me, it's that I have to be careful. And if there's one thing I've now got to be particularly careful about, it's the very man I've been assigned to monitor: Bickerbug. In fact, I'm so convinced of the need to be careful with him that I'm almost persuaded to take my assignment on him seriously. Here's what happened.

Word got to me on the morning of January 4th that he'd been released from detention the previous day. Naturally I got curious, so I decided to go and look him up. Lati pressed to go with me, but I managed to persuade her she might not like it – I almost told her she might not be welcome – and at any rate the man had just been released and might not be in the best of moods. So in the evening I drove over to Bickerbug's place at Obalende. But he wasn't at home. I knocked a couple of times and there was no reply. I turned the door knob, and to my surprise the door flew open. I almost thought the man hadn't returned to his place upon being released, but when I looked around the room I noticed a few changes. The books on the floor had been laid away in one corner, and the bed had been made. The floor had even been swept – well, not *swept* as such, but the reckless litter had gone. If I'd found Bickerbug at home I'd probably have commented on how salutary the detention must have been to him. I'm not sure how he'd have taken the joke, but I'd have tried.

Anyhow, when I didn't find him, I closed the door again and

walked across to his neighbours, the old couple who'd told me some three months ago how Bickerbug had been taken away by the NSS. This time, only the man was in. He received me more warmly than last time, and told me how happy they'd been when they saw Bickerbug back again. They thought he looked changed, a little weird, but they were happy nevertheless to welcome him back to the neighbourhood. He had thanked them for their concern, but told them he was all right and simply wanted to have a good sleep. So they'd left him. The next morning, however, when the man's wife went over and offered to give him breakfast, he'd gone – nowhere to be found! They had since then noticed that he came to the place only once, but didn't stay more than a few minutes, and then only in the late evening.

I thanked the man and left. I returned a few more times to Bickerbug's place within the week, but I never found him, and each time I found the room in almost the same order as before. It struck me there was something fishy going on. So one day last week I decided to try a different tactic. I took a taxi to the place. I knocked, opened the door and again there was no one in. So I shut the door and walked away, settling down at a spot where I wouldn't easily be noticed – on a pile of old, mouldy concrete blocks about forty yards from Bickerbug's place and on the other side of the street. This was at about 8 p.m. I'd been sitting there for about two hours, looking out for any sign of movement in the room, when I noticed a figure open the door, switch on the light and shut the door again.

Without delay, I got up, walked over to the door and knocked.

'Who's that?' his voice rang out at once from the room. It had got raspier.

'Hey, Ebika,' I said, 'it's me, Piriye. Can I come in?'

After a brief moment, he walked over to the door and opened it.

'Hello, my friend,' he said, beaming with a broad, toothy smile. 'How are you? Come on in and sit down. How are things?'

I couldn't believe this. He had on a black T-shirt and dark grey trousers, with a pair of sneakers on his feet. He was wearing those dark glasses. I sat down on his bed and took another look at the reformed Bickerbug, angry man suddenly turned into the most cheerful extrovert of all time. I hardly had time to return his salutation or say a word before he started on me again.

'So, my friend, how has the world been treating you lately?' he said, with the broad grin still in his face.

'Well ... okay,' I must have said. 'But where have you been, man?'

'Oh, here and there,' he said. 'You know, it's such a great thing to be free again. I go to Surulere to visit a few relatives and friends now and then. But I'm mostly around. Let me offer you something to drink. Er, what would you like to drink – beer, stout, what?'

'Nothing, man, nothing. I just wanted to ...'

'Oh, come on,' he interrupted me, still beaming broadly. 'You can't mean that. I'm sorry I have nothing in my room right now. But, er, let's ... let's go out to the bar over there and let me buy you something to drink. We must celebrate our freedom, don't you think?'

I was about to decline the offer, but he dashed quickly to me, held one hand across my mouth, and put a finger against his lips, indicating that I should keep quiet. I complied.

'So what do you say – shall we go over there and have a little drink?' he said, still grinning.

'Well, okay,' I said, playing along with I knew not what game.

'All right,' he said, clapping his hands cheerily. 'Let's go then.'

I got up from the bed, and as we walked out of the room he shut the door with a rather loud bang. Once again he held a finger to his lips. I was wondering what on earth was going on. Then he opened the door again and said, again very audibly, 'Oh, I've forgotten to take any money. Let me get a little change from under my pillow.'

He reached carefully under his bed, pulled out an object that looked rectangular and darkish, about the size of a pencil sharpener. He held it a little away from him and said, 'Okay, let's go.' Before I could move, he held me by the arm (I was still outside the door) and pulled me gently in. Then he shut the door with another loud bang. He turned over the object in his hand, moved a little switch under it, and said, 'There. That takes care of our friends.'

By this point the grin had completely disappeared from his face. He still had on his dark glasses. He took a deep sigh and flopped down on the bed.

I summoned the courage to ask, 'Now, would you mind telling

me what that was all about? What's come over you?'

In reply he opened his hand and said to me, 'Take a look at this stuff. What do you think it is?'

I studied the object, part metal part plasticoid, and didn't have to be told.

'I've never seen one before,' I said, 'but I suspect it's a bugging device.'

'Precisely,' he said, springing up from the bed. I followed him with my eyes as he moved to the window to take a peep out, then back again. One hand was in his pocket. The other one was loose but a little twitchy and so, I noticed, was his lower face beneath the dark shades. 'Precisely. Now you know what's come over me.'

'But why would they want to bug you?' I asked, playing my part well, though I hated every minute of it.

He stopped briefly in his roaming to look at me.

'You amaze me, Piriye,' he said. 'A man is held in detention for about three months. Suddenly he is released one day and told to go home, just like that. Wouldn't he have reason enough to be suspicious?'

'I know,' I said. 'But it happened to me too.'

'I'm coming to that,' he said, resuming his pacing, 'though you must realise you're not exactly in the same position as I am. You're just a nosy journalist, but me – I'm a marked man. There's a price on my head, and I have to be extra cautious. So the first thing I did when I got to my room was to take a good look around. A room that's been open for about three months, anything and anybody could have come in here. I searched every place thoroughly, shaking out every piece of paper and every strip of cloth. I finally turned my bed inside out, and there, stuck underneath the wooden boards against the head-post, was that thing in your hand.'

I took another look at the object, and there was indeed a needle or pin strong enough to stick into a wooden surface. Frankly, I didn't feel safe with that thing even though Bickerbug seemed to have turned it off. So I excused myself from the room and went out to put it somewhere safe in the backyard.

'Does it bother you?' he asked as I came in. 'I've switched it off.'

'I know,' I said. 'I just wanted to be on the safe side. You never know.'

'Oh, relax,' he said, smiling. 'Everything's under control. I know how far they can reach.'

I looked at him and shook my head. I couldn't help feeling sorry for him, and may have betrayed my sentiment.

'What makes you think they couldn't have bugged you even in the camp?' I asked.

'I thought about that too,' he said, reflectively. 'You know those two goons I had as bodyguards there?'

I nodded.

'Well, two days before I was set free they themselves were taken away. I suddenly felt disarmed. Not because I feared for my personal safety among the other inmates, but because I thought the enemy had made away with something more substantial than just physical security. I couldn't quite decide what it was, but I just knew I had to look out for myself more carefully than I'd ever done. And now, my friend,' he said, settling down again on the bed, his interlocked hands supporting his head on the pillow, 'tell me how they came to let you off. What happened?'

I gave a deep sigh. Now it was my turn to look out for myself.

'Well,' I said cautiously, contemplating the right words, 'some of us have our ways too, you know.' I turned to him with a smile on my face, thinking he would acknowledge the joke. But his face remained dead serious. 'Actually, it's my girlfriend. She has this uncle who's a judge, and he went around talking to some people, and that's how they got me out.'

'Judge,' he said, tightening his brow. 'What judge?'

'His name is Benson.'

'Benson,' he repeated, sitting up. 'Ekundayo Benson?'

'Yes,' I said. 'You know him?'

'Of the celebrated dock union case? Who wouldn't know Justice Ekundayo Benson?'

'What dock union case?'

'How could you have forgotten? The case in which twenty members of the dock union and their officials were sentenced to life in jail by the courts back in '68. Mohammed had them released in '75 – that was one of the first things he did that endeared him to the people. You call yourself a journalist, and you don't know about all this?'

'Well,' I said, a little humbled, 'that was a long time ago. In '68 I was just over a year old in journalism. But I confess you do have a good sense of history.'

'You would too if you were in my position. Well, well, well, Piriye,' he sighed, resting his back on the bed again, his hands once more interlocked under his head, 'I must say you are getting on in the world.'

'How do you mean?'

'Well, you're in the big league now. Er ... this girlfriend of yours – she's the girl at the *Chronicle*, isn't she?'

I said yes. I turned to look at him, and I could feel his eyes penetrating me through his dark shades. He was nodding his head slowly and humming 'M-hm. M-hm.' I couldn't figure out what was on his mind, but I began to wonder if he still trusted me. Had I now created a distance between us? Had I stirred ethnic sentiments within him? I couldn't stand his staring any longer, so I looked away.

'Well,' I said, 'you know what happened between my wife and me. I couldn't help ...'

'Come on, Piriye,' he interrupted, sitting up and putting a hand on my shoulder, 'you don't have to explain anything. I'm not a child. I know what you've been through, and I wouldn't blame you. Besides, I like Miss Ogedengbe. She's a fine girl. She's warm-blooded, and she has guts. She's my kind of girl. I'm sure she'll be good for you. Hell, you have guts too, coming all the way to the lion's den to ask about me! You'll make a fine pair,' he concluded, giving me a cheery slap on my back and falling back again on the bed.

I appreciated the gesture, and I felt reassured.

'Anyway, where have you been, man?' I asked, changing the subject. 'As soon as I heard you were out, I came here looking for you. I kept checking on you for over two weeks. I even asked your neighbours, and they admitted they'd seen you once and welcomed you, but hardly saw you again afterwards.'

'I know,' he said. 'I don't trust them any more. Man, after what I've been through I decided I had to be careful who I mixed with. How do I know they're not friends with the NSS?'

'Maybe,' I said, wincing within myself and being careful not to let him sense anything. 'So where have you been all this time, man?'

'Oh, here and there,' he said, rising from the bed and pacing again, with a brief visit to the window for a look out. 'Here and there, you know. Hey,' he turned to me, suddenly bubbling with life, with a smile on his face, 'you can't imagine what an exciting time I've been having, educating myself. Here.'

He fished among the books in the corner and brought out about four or five volumes. One was a Civil Engineering book I'd seen before in the room, but the others were on other subjects – petroleum engineering, a book on dams and bridges, another on environmental pollution and another on petroleum law. I was a little perplexed.

'Engineering, pollution, law,' I said, flipping through the volumes. 'What are you doing with all this stuff? I thought you were an English teacher. What's all this got to do with who killed Cock Robin?'

He chuckled briefly.

'Man,' he said, 'we live in an exciting country, and these are exciting times. There's so much happening in this country right now, I just have to educate myself so I can find out how to deal with the problem.'

'What problem?' I asked – rather naively, I must admit.

He stopped pacing to look at me, then shook his head and continued pacing.

'Don't be ridiculous, Piriye,' he said. 'I have been to jail, you have been to jail, and you still don't think there's a problem?'

'Okay. Okay,' I said.

'I've been around, man. Travelling round the country, talking to people, finding out about things, reading books. Look,' and he sat down on the bed, close to me, 'we've got all these problems about the Kwarafa Dam cutting short the water flowing down the Niger and messing up the fishing economy of our people. Now what do you know about dams and how they work?'

'Nothing,' I said.

'There you are – so how can you deal with the problem of dams if you don't know anything about dams? And we talk about the oil pollution in the Delta, and how that too has been destroying the fishing and the farming life down there. How much do you know about oil pollution?'

'Very little.'

'There you are. So that's why I've decided to educate myself – travelling around, talking to people, finding out about things, reading all these books. Hey,' he turned again to me, bubbling once again with excitement, 'you can't imagine what an exciting subject Engineering is – I wonder why I wasted all those years reading English. Shakespeare, Conrad, Joyce, Achebe – all that nonsense.'

'Everything counts,' I said.

'I know,' he said, 'I know. But Engineering is out of this world. Like – what do you know about bridges? I mean, about the history and the mechanics of bridges?'

'Nothing.' I was observing him with interest, as he launched into his erudite lecture, which I can only *try* to reconstruct here.

'Do you know that *we* – our people, I mean – we taught the white man how to build bridges? That's right. Our people, the Third World – we were building bridges hundreds of years before the white man caught on to that technology. Admittedly, our ancestors built them from creepers and twisted rope cables held between posts made of hewn timber or tree trunks. But the principles of these suspension bridges, which allowed our people to make their way across gorges and ravines, had been laid down long before the white man came in with further refinements. The Incas of South America built suspension bridges from vine cables tied to timber beams held between rocks, and with these they were able to cross huge canyons. Early forms of wire suspension bridges are still to be found today in places like Sweden and the Scottish Highlands, but the Chinese were the first to build metal suspension bridges way back in the early seventeenth century, during the Ming dynasty. Ropes were by then replaced by link chains, and the tower beams were of solid masonry. Of course the Turks and the Romans built stone arch bridges way way back, and their remains can still be seen today. But the Chinese made the first modern, I mean metal, bridges. All the wonders you hear of today – Finley's Merrimac Bridge in Massachussetts, Roebling's Niagara Falls Bridge and Brooklyn Bridge, even the Golden Gate Bridge in San Francisco and the one at New York Harbour – all these were simply improvements of the Chinese model.'

He got up again, did a little jig and resumed pacing.

'Interesting,' I said.

I was genuinely impressed by all this learning. I leaned back against the wall to study our friend. Believe me, Tonwe, the way he was pacing about and gesticulating with his hands, he looked like a man possessed. I could almost say he looked like a university professor, but for his dark spectacles and his dark T-shirt and dark grey trousers, which gave him something of an eerie, ferocious aspect. Frankly, I was becoming genuinely mystified by him.

'Do you know any of the factors underlying the construction of modern bridges, the kind of forces and pressures acting on the total structure which have to be reckoned with in the making of them – like the dead load and the live load of a bridge, the wind load, the dynamic effect of vehicles moving on it, and other forces such as erosion, land slides, earthquakes, stream flow, et cetera, et cetera. Do you know about all this?'

I shook my head.

'Well, it's a whole new world of information, man,' he said. 'We see all these things and we take them for granted and we don't ask any questions until the government and their agents and collaborators start messing with our lives. Now, what do you know about the building and installation of oil rigs?'

'Nothing,' I said.

'But you should, you know. You should,' he said. 'You're a journalist investigating these things. Besides, the oil exploration is the biggest problem of our people down in the Delta, so how can you understand the problem if you don't study it? You know, Piriye, it was interesting reading about the Warri meeting of the Task Force on pollution. I admired the contributions made by your old colleague, Mr Brisibe. Although the papers, as usual, did not carry a detailed report of the proceedings, I've been able to get a full transcript of the discussion that went on in that NOC Conference Room.'

At that I sat upright, consternation written all over my face.

'Relax,' he said, chuckling. 'I've told you – in this struggle you've got to ...'

'Stay one step ahead of the chase,' I chorused with him.

'Right! Look,' he said, squatting to conduct his lecture with more concentration. 'There are many kinds of oil rigs, but every one of them is prefabricated on-shore, or assembled from component parts, as here in Nigeria, and transported to the off-shore drilling site on a barge. There are two main parts to the rig. First, there's the substructure which provides the stable base for the drilling operations, and then there's the deck on which the entire exploration tackle is mounted. The substructure itself is of two kinds – the steel template kind and the gravity kind. The steel template structure rests on some four or eight legs lowered to the sea-floor and secured to the sea-bed by pipe piles driven some two to three hundred feet below the sea-floor. The gravity structure is used in places where the sea-floor is too hard to be bored by pipe piles, as in icy Alaska, or where there is some rock in the area that may prove to be an obstacle. So the gravity structure consists of heavy concrete cylinders – metal cylinders have also been used, even here in Nigerian off-shore rigs. These are lowered to the sea-floor, sixteen of them or so, holding in place the three or four other cylinders on which the operations deck is going to rest. It just sits on the sea-floor. The deck itself must be sufficiently high above the water not to be buffeted by waves. *Then* the well-head, from which the borehole is going to be sunk, is fitted on a cellar-deck about fifteen feet below the main or operations deck. You must understand that all this rigging is a massive structure – what you see rising from the waters like one huge Christmas tree or mast is a mighty bulk, you know.'

'Is that right?'

'Oh, yes. A standard operations deck is some two million pounds in weight, while the substructure – because it has to withstand the various environmental forces or loadings like wind, current, wave, even seismic action – is usually twice that weight. In fact, the gravity structure may be as much as ten or twenty times the weight of the steel template, and can therefore carry a correspondingly heavier operations deck. Man, those rigs you see are out of this world,' he said, rising now and resuming his pacing.

'Interesting,' I said. 'Simply amazing.'

'*Then* comes the oil drilling itself. I'm not going to go into the mechanics of exploration and transportation of the crude, as this may take us all night. I'll simply concentrate on those hazards and the resultant pollution which has made life virtually unlivable for our people down in the Delta and which the Frank Segals and the Artauds and Cioffis of this world will never be honest enough to own up in an open forum. Your man Brisibe was on the right track of questioning, but he didn't go far enough.'

'Obviously,' I cut in, 'he hasn't given the problem as much attention and study as you have.'

'I know. Now let's look at the various sources of pollution. First, the well-head is fitted with what's known as a blowout preventer, which shuts in and controls the well in case the oil or gas issuing from the drill exerts a pressure higher than that exerted by the column of drilling mud in the borehole. But the preventer is never able to stop a blowout, because there is as yet no foolproof device for balancing one pressure against the other. The only effort made so far is to reduce the volume of oil blown out, from some one or two thousand tonnes to about six hundred tonnes per rig.'

'So there's always an overspill?'

'There's always an overspill – you've got it. And when I talk about a blowout, it's really a *blowout*, an explosion. The rig can take it, because it's got the weight to absorb the shock. But what about the villages in the environs? For them it's another tremor, and this goes on constantly even before the oil drifts to their fishing enclaves and their farms.'

'I see,' I said. 'I see.'

'All right, but that's not all. Overspills occur from various other sources. The pipelines taking the crude from the well to the reservoirs and barges may spring a leak or get fractured due to a variety of causes, and of course the oil escapes into the water. Or the barges and bunkers carrying the oil may lose some of it for various reasons – leakage, overloading, blown gaskets, even corrosion from prolonged chemical action of brine and other things – and again all that oil is emptied into the water. And that's not all.'

He stopped to sneeze and blow his nose. He was becoming increasingly agitated, and didn't even wipe his nose properly before continuing with his lecture, which I must confess I was enjoying.

'Every once in a while the oil tankers are washed – and that's another source of pollution. The method is called *ballasting*. Sea water is taken into the tanker to clean out the oil sticking to the insides of the hold. The sea water is the *ballast*, and after the tank has been cleansed it is *de-ballasted* – that is, the ballast and oil are thrown out into the surrounding water. Now, an alternative to de-ballasting is what's called the Load-on-Top, or LOT for short. By this method, the ballasted residual oil is retained in the tank, and later the ballast is strained off, allowing the residual oil to settle at the bottom of the tank while the next cargo of oil is loaded on top of it. But this is very seldom practised, because the tanker may be loaded with a different type or grade of oil from the one it previously held. So de-ballasting goes on all the time, and as the volume of exploration from the various oil companies increases the volume of oil pollution increases also.'

'I should imagine it does,' I said.

'Okay. Now, the dangers of all this oil pollution to the environment are sufficiently well known to you. The fishes die because the floating oil blocks the oxygen from the water or because their respiratory membranes are clogged by the oil. Even the birds that dip in the water to catch fish and other foods suffer – their wings are matted by the oil and they cannot fly so they sink and drown or die on dry land from asphyxiation, having taken in so much grease. The farms, too, are ruined – the crops won't grow because the oil floating on the irrigation chokes the soil. Even the drinking water is affected – your friend Brisibe was right, and Chief Zuokumor was only a paid fool. What else did you expect him to say? In fact, one oil executive once told me here in Lagos that he didn't know what all the fuss over the drinking water was about, for after all it was only like taking in water with a little kerosene in it. When I asked him if he would like to drink water with kerosene on a regular basis, he said he would if he had no choice! Can you beat that? But the truth is that drinking water so contaminated causes various forms of enteritis, some more severe than others. And you know why the truth is never told to the people down there?'

'Because some of our officials collude with the oil companies to suppress it,' I replied.

'You're damn right,' he said, stabbing his index finger towards me to stress his point.

I could see spots of sweat on his brow and on his nose. He was pacing with increased speed.

'Now,' he continued, 'what have the oil companies done to avoid all this pollution or at least reduce it to an acceptable level? Nothing. Of course, there are various international laws for ensuring the safety of the environment from severe pollution and also for enforcing standards for oil exploration and conveyance. But when national officials turn a blind eye to all the risks, the oil companies have all the freedom in the world to ignore these safeguards. Now –,' he cleared his throat before continuing, 'in other parts of the world there are methods used for ridding the oil pollution. Segal talks about his company's plans for using the most advanced detergent technology, as if that's something new. What he's really talking about is *dispersants*, some kind of chemical mixture of a detergent emulsifier and a hydrocarbon solvent which is sprayed on the floating oil by tugboats specially designed for the job. The quantity of dispersant sprayed depends on the type of oil spilled and the thickness with which it floats on the water. The mixture of dispersant and oil is then stirred by the skews – some of them look like giant blenders – fitted to the tugboats, or by some other kind of agitators towed behind them.'

'Wonderful,' I said. 'So there are huge amounts of foam formed by all this?'

'Exactly,' he said. 'You know – it's like stirring oily water with Surf in a basin. And don't forget – these chemicals are extremely toxic, far more so than Surf or anything like that.'

I nodded.

'Okay,' he continued, suddenly clasping his arms around him as though he felt cold. He seemed in fact to be shivering a little, and yet he still had those spots of sweat on his brow and his nose. 'Now, there's an alternative to dispersants, and it's called the *Vikoma Seapack* system. Here, a ship releases a boom that will push the slick downwind to a safe corner of the shoreline, and then collects it through a suction pipe into a waiting tanker. But these various methods – dispersants, Vikoma and others – they all have their problems. Take dispersants. Again, your man Brisibe was right about the hazards of detergents to the environment, and Segal was only being dishonest and vicious, reducing the whole thing to a joke. The simple truth is that these detergents are frightfully toxic. They could kill off all the fishes in the creeks just as swiftly as the slicks do, perhaps even more swiftly. And that's why their use is recommended for large open sea locations. Segal was only fooling. He knows all this, but even more than that he knows that his company has no *real* plans to use the detergents. The same goes for the Vikoma Seapack. These devices are really very expensive, and in the face of the fierce competition between the ten or more oil companies operating in this country, none of them is prepared to commit any substantial outlays to combating pollution when they are not sure that these can be conveniently offset by returns from their explorations. In fact the books will tell you that in spite of the various methods available for ridding the water of oil pollution, there is often no alternative to the payment of damages to the affected communities through arrangements set up for adjusting claims. But what do our people get for their claims? Every time they complain they are called all sorts of names – troublemakers, saboteurs, greedy and rapacious – and harassed by government officials paid by the companies to keep our people from bothering them. You see, there's no way we can win. So what's there to do but speak to them in the only language they can understand?'

He stopped pacing to look at me. I couldn't stand the concentration of his gaze, though he still had on his dark glasses. I shook my head and looked down, indicating I was just as baffled by the hopelessness of the case as he appeared to be.

He continued pacing, his arms clasped across his chest, and he was still shivering. I must say that, this being the harmattan season, the nights are usually pretty cold. And by this time it was getting quite late – I looked at my watch and it said something like 10.30. Bickerbug moved to one of the two windows and shut it, saying apologetically that he felt rather cold. As I said, these are harmattan days, and I should also add that the T-shirt he had on was the sleeveless kind. Still, it didn't seem *that* cold to me, and I would have thought that a man who'd been lecturing with so much enthusiasm would indeed welcome the cool air.

Anyway, as he moved to shut the window I sighed deeply to indicate I was a little tired and that it was about time I was on my way home. Apart from anything, Lagos is becoming increasingly unsafe for late nights out, and I had come without my car. So I was about to tell him I had to be going, when suddenly it seemed as if he was only just beginning his lecture.

'And what do you know about the great Kwarafa Dam, the one they've told us is going to solve all our industrialisation problems?' he asked, his arms still clasped around him.

'Nothing,' I said. 'Listen, man, I really have to be ...'

'All right, all right,' he cut in. 'I won't go into the mechanics of it, although I've read all the available project reports and went down there myself recently. But there was something Segal said in that transcript I read which really made me mad, because he knows the truth and isn't honest enough to admit it. When Brisibe asked him if the detergent chemical wouldn't kill off the fishes or poison them and the water they live in, Segal answered that the waters are constantly flowing down to the sea, so the chemicals won't linger long enough to cause any damage. But that's the whole fallacy about the dam. We've been told – the books even say so – that the advantage of a dam is that it controls the flow of water in a river so that the river doesn't flood in the event of heavy rains, and all that garbage. But you know the *real* problem there, don't you?'

I shook my head.

'The real problem there is that those who planned the construction of the dam never took into consideration the possibility of oil being found in the Delta and whether certain ecological problems might arise due to the restriction of the water flow. I'm not talking about the reduction of the fishing activity – what else creates an abundance of fishes but an abundance of water? So in reducing the volume of water they have depressed the fishing economy. But I'm not talking about that. I'm talking instead about upsetting the ecological balance between the flow of water from the river down the creeks to the sea and the periodic surge of the tides of the sea up the creeks. The force of the flow downstream is now weaker than the surge of the sea tides upstream. There are two consequences of this. First, there is greater salinity in the waters of the creeks now than there used to be in the past. This is all right for the oil companies, because they no longer have to go far out to sea for the ballast needed to clean up their tankers – they can simply clean them close to the creeks and throw the sludge into the waters right there. It's bad for our people not only because of the sludge but also because the volume of natural fresh water from the river is now severely reduced. But even more serious is the fact that, with the reduced force of flow down to the sea, the pollution lingers far longer than it usually would. So that when Frank Segal says that the detergent chemicals aren't going to be there long enough to do any harm to the fishes and the farms, he's simply talking bullshit. He knows it, but he'll never tell the truth, and he's got enough people in this country to help him keep up the lie. Man, our people down there are in an awful mess, and there's no one prepared to do anything to help them. I tell you, Piriye, *something has got to be done*!'

He unfolded his arms long enough to stress that point with a clenched fist, but put them back together again soon after, grinding his teeth now in a clear indication that the cold was really starting to get to him.

'But you know something, Piriye?' he said, as I made to get up.

'What?' I asked, my hands bracing the edge of the bed.

'If you think I'm the only one who feels this strongly about the situation in this country, man, you're in for a big shock. There are so many angry people in this country, so many people with so many grudges – it's a miracle the country is still in one piece. I know, man. I've moved around and I can tell you: this country of ours, this big and mighty country of ours, this fucking giant of Africa as she loves to hear herself called, this glorious asshole of a nation – she is sitting on a time-bomb, and it's only a matter of time before it ticks to the appointed time and the explosion lights up our faces. Do you know who's angry?'

I shook my head.

'I'll mention just a few,' he said, stopping to spread his fingers for a count-off. 'First of all, our people down in the Delta are angry because too many promises have been made to them – compensations to be paid, roads to be built, et cetera, et cetera – and nothing ever gets done. You've heard of riots down around the oil installations, so that's not new. The Nigerian executives at the rigs – engineers and technicians – they're angry, *very* angry, because the white rednecks are paid far more than they even though the Nigerians are better qualified. The small workers are angry – across the country, of course, but even more so the workers in some of the key services at the rigs, the power stations, the dam, all sorts of services – because their overtime claims are piling up and their conditions of living are intolerable, yet nobody seems to be listening to them. Everybody calls them greedy. But listen to this. Even the *white* men in this country are angry for various reasons. You won't believe it! The British are angry because the Federal Government sidestepped them and gave the contract for the Mainland Bridge to the Germans. The Russians are angry because the contract for the steel company at Aladja was given to the Germans when they could have taken that along with the Ajaokuta steel project. The Israelis are angry because when, like other African countries, we broke diplomatic relations with them over the Six Day War, we terminated several lucrative contracts we had lined up for them. Even the Germans are angry because ... look I'll show you something I received just the other day.'

He stooped down by the pile of books at the corner near the

bed, and as he did so something dropped out of the right pocket of his trousers. It was a little packet of something. I tried to pick it up for him, but he was quick to notice my movement.

'Don't touch it!' he barked at me, snatching up the packet.

I was shocked beyond words. He was looking fierce and furious, like a cornered animal.

'Come on, Ebika,' I said, spreading out my arms in a gesture of harmlessness. 'I was only –'

'Don't touch me!' he barked again.

'Come on, don't get so excited. I –'

'Don't get excited, he tells me! What do you know about getting excited? Do you know what I have been through?'

'What about me?' I retorted.

'Yeah, you! What about you? You were there for barely two days. But I was there *three* months – three *fucking* months! And do you know what they did to me down there? Hm? Do you know what they did to me? Take a look at this!'

He snatched the dark shades off his eyes and – God, Tonwe! I couldn't believe what I saw. There were dark brown patches and blotches around his eyes, a welt below the left brow and a scar across the bridge of his nose, between his eyes. I had never seen all these throughout my two days in the detention camp or at any time before now, because they had been concealed by the broad dark glasses. I swear, Tonwe, that boy must have been hurt very badly by somebody. Even the whites of his eyes had a thin film of red over them, indicating, I think, that they were still recovering from some harsh encounter. To be honest with you, I was a little frightened.

'Good God, Ebika,' I found myself exclaiming softly, 'what on earth happened to you?'

'What happened to me!' he mocked. 'Hey, you haven't seen a thing yet.'

He pulled off the black T-shirt, spread out his arms and turned his body slowly round for me to see. Welts, slashes and cuts, only recently healed, were all over his body. These horrible dark brown marks were all too visible on his light mulatto skin. I tell you, they filled me more with horror than with pity. At this point I was simply speechless, and merely looked on with a hollow mouth as he pulled the shirt back on and fitted the dark shades over his eyes once more. His face had begun to sweat even more.

'Have you seen enough now?' he asked.

His face had gone twitchy, and his already raspy voice was further roughened by an all too obvious effort at self control.

'Man, what did they do to you down there?' I managed to ask.

He sat down slowly on the bed, trying to calm himself. With his elbows on his knees, he held his head between his hands and seemed to ignore me for a while, shaking his head lightly. I was still a little too frightened to repeat the question or make any kind of move. Then he brought down his hands and raised his head slowly.

'I was led handcuffed to the gate of the camp,' he began after a deep sigh. 'Yelwa made sure he saw me to the gate. While it was being opened, he shouted to the armed guards in the camp, all four of them. When they came he told them to "break" me and leave me in the "box". Two of them led me away, while the others cocked their guns in case the other detainees, who were now beginning to gather around us, tried anything.'

He stopped briefly to clear his throat and wipe his nose, which was running a little.

'They took me to the back of the building, and there they worked me over. The two armed soldiers drove the crowd of inmates some distance away, to give themselves enough room for their business. While they held their guns at the ready, facing the crowd, the two others loosened their belts and began to lash me furiously. Remember, I was handcuffed, and could not ward off the lashes in any way. I screamed at the top of my voice as the pain of their blows cut through my skin. After a time, I was thoroughly weakened by the pain, and I fell down. They were saying certain things as they lashed, but I can't recall what it was they were saying. I just know they were not singing my praises. When they got tired of lashing, they resorted to punching and kicking – on my face, my sides, my balls, everywhere. I don't know what else they did, because I passed out. When I woke up, in the middle of the night, I found I was lying on the floor of a narrow, windowless room. I could tell the shape of the room by a faint light that seeped through the cracks in the door. I noticed my handcuffs were gone. I was lying on a wet, stinking floor, and I was aching all over. I could *not* get up, however hard I tried. In the morning I saw that there were faeces and urine all over the floor. Mind you, I could *barely* see, with my eyes all swollen and bloodshot and red, and the daylight faint. But I could tell that this was the "box", a cell for the solitary confinement of those who had earned the special resentment of the security police and so deserved to be taught a good lesson. I couldn't have eaten even if they'd brought me food there – man, you never smelt such a stench in all your life!'

He shook his head, spat on the floor and rubbed away the sputum with the sole of his sneaker.

'Terrible!' I muttered. 'Terrible!'

'Yes,' he nodded. 'They certainly taught me a lesson there. That whole day passed, and then the second night too, and I still wasn't given anything to eat, nor was I allowed even a moment's exposure to the open air. It wasn't until the mid-morning of the third day that I heard the lock turning and saw the door open wide. A guard shouted my name and told me to come out. I crawled my way out, weak and thoroughly broken. I felt my eyes and they were swollen. My waist was aching. My knees could hardly support me. There was blood and shit all over my clothes. I was stinking, man, and I was *starving.* The guard led me to the back of the building, close to the wall. He threw a cake of soda at me and turned a hose of water on me. Somehow, it was a great relief. Although I was weak, I did the best I could to scrub myself all over and rub the stains and the filth off my body. I was wet, but I felt fresh and clean. The guard took me to a regular cell and gave me something to eat. I can't remember what it was or how it tasted – I think I gobbled it too fast to realise what it was I was eating. Hm.'

He shook his head and smiled, then held it again between his hands. I didn't know what to say.

'So please don't tell me about getting excited. I've had my share of excitement.'

'I'm sorry, man,' I said. 'I didn't know anything at all about this.'

'That's all right,' he said. 'It's not your fault. I know who my real friends are,' and he volunteered a smile, 'and they know me. But mark my words, Piriye. They will pay, this country will pay for every single blow and every cut they gave me in that camp. They will pay, man, and pay dearly.'

'What do you mean?' I asked.

'That's all I can tell you. Remember it, and don't forget to put it down in your book exactly as I have said it. Every single blow and every single cut – they will pay dearly for it.'

He got up from the bed and began pacing again, his hands clasped around his body once more. I could feel the closeness dissolving between us. He probably didn't mean any offence. But he'd suffered such a deep personal injury that he couldn't find words strong enough to articulate the bitterness – my question probably sounded like something of an intrusion into that private agony. I didn't feel like pressing him. Besides, it was close to midnight by this time, and I was fearful of my journey home, considering the nature of Lagos night life these days. I sighed deeply and got up from the bed.

'Ebika,' I said, 'I've got to be going. I'll see you again soon.'

I noticed he was shivering again, as he nodded his head to accept I had to go. Somehow I felt a bit sorry for him. Despite the stubborn self-assurance he has constantly projected, I wondered what the future held for him or how he hoped to survive now that he had lost his job at the school and was in trouble with the authorities. As I stepped towards the door, I felt in my pockets and fished out some money. Twenty-five naira. I thrust N20 of it into his folded arms and walked away, not certain how he was going to treat the offer. It was a relief to hear his voice.

'Hey, Piriye, I really appreciate this. Hey, I'm sorry about my little tantrum tonight. I never meant to do it. And, listen, thanks for this.'

'Forget it,' I said. 'You might do the same for me some day.'

He held me by the shoulder as I stepped outside the door. There was a somewhat conspiratorial feel to his hand.

'Look, er ...' he said, hesitatingly. 'As you can see, this place isn't really safe for me any more. I'm keeping it, but er ... that's only to keep them at bay. I'll tell you where you can find me, and when. How well do you know Ajegunle?'

'Not well,' I said, 'but I can find out.'

'Good. Just look for Fatigbe Close. No 8. Ajegunle isn't an easy place for me to direct you to, so you'll have to find your own way. Fatigbe Close ends in a junkyard of broken down trucks, beyond which there is a swamp. No 8. You'll find me there most days between 12 noon and 6 p.m. Okay?'

I nodded. 'Thanks,' I said, 'I'll remember.'

'Hey,' he said, gripping me again on the shoulder, 'thanks

again, Piriye. You've put yourself out for me so much, I'll never forget that. You are a real brother.'

I patted his side in turn, and walked off.

'Hey,' he called again. 'Don't forget our little friend you hid away.'

We both laughed. I went to the backyard and brought back the bugging device. He took it from me and lodged it very gently in its usual place under his bed. I had stepped into the room, but he waved me back and motioned for me to shut the door. I did so and stood looking at him from outside the window. He switched on the device, and tiptoed carefully to the door. Then he made an elaborate show of unlocking the door, and swung it open.

'Oh, what a night,' he said, laughing boisterously. 'Boy, I'm so tired and I'm stuffed. What time is it?'

'Oh, about eleven o'clock,' I said, playing along with his mock cheerfulness.

'Boy, that's late, isn't it?'

'Yes, I'm afraid I've got to be going home. So I'll see you some other time.'

'Okay, man, take care.'

'Goodnight.'

'Goodnight.'

I waved at him. We exchanged conspiratorial smiles, he shut the door with a loud bang, and I walked away.

Tonwe, I have to admit that my meeting that night with Bickerbug was one of the most informative I've ever had. All those academic details about oil exploration and oil pollution were thoroughly enlightening. I must read up on them to supplement the information given by Bickerbug, for certainly they will come in handy when we come to do our book on this whole problem. I know we are going to have to talk to the oil companies and various officials to get their own sides of the story. But I must confess that Bickerbug's revelations opened up my eyes more than a little.

Despite that, I'm a little worried. I told you at the start of this letter that I've got to be careful about him more than ever before, and I meant that seriously. I still don't think I'm any less angry than he is about what's happening to our people down there in the Delta, except that he has, probably for good reasons, taken the matter more personally than I've been inclined to. Of course, I too have suffered unjustly in the course of my investigations, and I will be content to lead my friends at the NSS a merry dance if only to get even with them for messing me up the way they did. But I'm not sure that I'll go along with the violence that Bickerbug seems to be planning.

I don't know – I can't be certain what that little packet was that he was desperately struggling to hide from me that night, but I'm pretty convinced it's something that wouldn't do him or anybody else much good, considering how violently he reacted when I tried to pick it up for him. And for him to swear that he is going to teach this whole country a lesson, just because the over-zealous men at the NSS did him physical harm – I don't think I want to be part of any such plan. I am worried because, although circumstances have brought me so close to him that I am inclined to consider him a friend, and have even gone so far as to describe him to the NSS as one, I am no longer sure I can fold my arms and watch while he plans an evil that may destroy everything and everyone, including the rest of us. I may not turn him over to the NSS, but I am almost certain I'm going to stand in his way if I see him make a dangerous move. I remember you saying that the Civil War was one trauma too many for this country. I think I agree with you more now than I may have done at the time you said it.

You may wonder why I seem to have done an about-face. Well, everybody learns from experience, and I'm no exception. This country belong to all of us, whether we like it or not, and it would be a shame to lend any form of support to those working to bring it down. Now, I'd be a fool to ignore the harm that ethnic prejudice has done and continues to do to interpersonal relations in this country. And yes, I'll never forget the evil that Ajibade and his like did to us in getting us retired from the *Chronicle*. But life with Lati has taught me to be careful not to think that all Ibile people are like Ajibade, with no feeling in their hearts beyond vicious self-interest and ethnic prejudice. Lati loves me and I love her and, though I am as Beniotu as anybody can be, where she comes from no longer matters to me any more than where I come from has ever mattered to her.

Besides – and perhaps more importantly than anything else – she is going to have our child! If the present generation of this country, in the blindness of its greed, has done little but promote mutual intolerance and endanger the life of the society, it nevertheless has no right, we have no right, to persist in our folly to such an extent as to destroy every chance our children may have of saving her. As I said, I'm in no hurry to turn Bickerbug over to the NSS and I would like to continue to treat him as a friend and even as a brother, as he has called me. But he does not have my support for whatever action he seems to be planning, and I will be more than ready to prevent him from ruining the chance that the rest of us have of living in peace and happiness.

Incidentally, Lati and I have decided that we will very shortly announce to her parents and relations our plans to get married. We're going to have a traditional wedding first, since I need time to settle the business over my separation from my first wife – three years and all that. But as soon as that's over, I'll legalise my union with Lati and make her a fully respectable wife. Meanwhile, we'll have the traditional ceremony so as, at least, to give the traditional legitimacy to the child we are eagerly expecting. I will let you know when we have fixed the date for the ceremony and will be more than happy if you can find time to be present, although I fully appreciate how rarely you entertain the thought of journeying far from your village.

I also hope you get onto the monitoring committee. In spite of appearances, Batowei has no choice but to redeem his ignoble showing at the Warri meeting and live up to his commitment to you. I am eager to know how that business turns out. On my part, I will seek out Bickerbug in his hideout at Ajegunle, a place about which I've heard the most scary tales but have never really got to know. I'm glad I don't have to be there after dark, but his choice of the place as his refuge from the NSS makes me even more curious about his plans.

Regards to madam from Lati and me. If she could see our growing collection of baby clothes!

Yours very sincerely,
Piriye

Brisibe Compound
Seiama.

8th February, 1978

Dear Piriye:

I found your letter very moving, especially the latter part of it where you talk about living in peace and happiness, and the future of our country. I should perhaps thank you for paying me the compliment of citing my own arguments in support of your own sentiments. But I too have learned from experience. I came home to live in peace and whatever happiness is possible, given our meagre means. But I have grown increasingly doubtful that the conditions for these exist here. Indeed, I am almost convinced that this country has little interest in creating those conditions or encouraging their growth amongst us.

Consider these facts. Fact number one: the monitoring committee has been set up, and I am not on it. My son-in-law at Warri did a bit of nosing around, and came here last weekend (the same day, in fact, that Priboye brought your letter to me) to give me the news. As God is my witness, I am not worried that I am not a member of the committee. After all, I never claimed to have all the wisdom in the world, or all the answers to our problems. There are much better men around than I.

But who are the three wise men who make up the committee? I am sure you can guess: Commander Adetunji is Chairman, with Mr Oluremi Ketiku of the Warri Port Authority and my good friend Chief Bieye Zuokumor as members! Not being one of them, I cannot tell you what are their terms of reference, nor exactly how long they are supposed to be doing their monitoring job. As I said, I should have little cause to complain that I have been excluded from the committee, since I cannot claim to be one of the wisest men in this community or area. But I do question the composition of this particular committee, and I have doubts about its intentions. Zuokumor is well known in these parts for his corruption. It is clear that the problems of our people will be of little concern to him so long as his personal interests are adequately served. I have no doubt in my mind that the Frank Segals of the oil business will rejoice at his inclusion in the committee (that is, if they did not directly engineer the decision).

I am equally disturbed at the presence of Commander Adetunji in the whole affair. After what Mr Opene and I experienced at the hands of the officer in September last year, and his clearly dubious role at the meeting of the Task Force in Warri, I need hardly be told that he had a hand in my exclusion from the committee, or which way his sympathies in this matter lie. I do not get much comfort from the presence of Mr Ketiku either. It may be argued that, as public officers charged with coastal duties, there was no way that he and the navy commander would have been left out of the committee. But Mr Etim Essien of the NOC, who is himself a public servant from a government parastatal, could just as well have been chosen for the job. Not only is he an oil engineer, but as a man from the Cross River State he is much better placed to appreciate the peculiar problems of riverine communities like ours, than are Ibiles like Adetunji and Ketiku.

Please understand that I have nothing against the Ibile as a race, and that I perfectly sympathise with the sentiments you expressed about them in your last despatch. But having been hounded out of Lagos by the Ajibades of this country, I have no wish to be further tormented by the likes of him in my own homestead. Nor do I find it funny that they should use men of questionable character from among our people to give a pretence of objectivity to their corrupt schemes.

This brings me to fact number two. Perhaps I should more correctly title it as a proposition, since I do not yet have any incontrovertible proof: my trusted ally, Mr Opene, who I should think has seen worse times in this whole palaver than I have, seems to have joined the other side after all! My suspicions first arose when I visited him a few days after I returned from the Warri meeting. I had gone to tell him how the meeting went, and to assure him that all was not lost (as I hoped and still do). He was not at home when I called. His wife told me he had "travelled". I thought this was odd. We had become sufficiently close, since events brought us together, for me to think he would let me know if he planned any major movements. But what I found even more strange was the changed appearance of his circumstances. There were two brand new fishing nets drying in the backyard, through which I had come into his house. But these were nothing compared to the brand new furniture I found in the house. For a fisherman, a new fishing tackle may not be such a startling revelation. But brand new upholstered furniture, for a man of such meagre resources and taste!

I swallowed my surprise. Before I left I urged Mrs Opene to ask her husband to look me up as soon as he returned. Not only has he not done this yet, but just a week ago, as I was paddling home from a fishing trip, an engine boat sped past me and tossed my little canoe about in its waves. In it was my dear friend Mr Opene, smartly dressed and ensconced near the Captain whose identity I could not easily establish!

Many letters ago, I pointed out to you that Opene had lived so long among our people, and identified himself so fully with our fortunes, that it no longer crossed anyone's mind that he was not a native-born Beniotu man. Piriye, I would be lying to you if, as I have watched events unfold before me in the last few weeks, I should tell you that I feel the same way about Opene as I did when I took him to see the navy commander in Warri. If those who have set themselves against the welfare of our people have not found it difficult to co-opt native sons like Zuokumor onto their schemes, you can imagine how much easier they would find it to drive the same wedge between us and men like Opene who have no roots here but happen to enjoy a position of esteem among us. As I said, I am not firmly certain of the facts. I have not sufficiently recovered from the shock to ask further questions. I hope to God I am wrong. But I rather doubt it.

I should welcome the opportunity of being present at the traditional ceremony uniting you with Miss Ogedengbe. But in the present state of affairs, I hardly think I am in the best disposition for such an event. I can only wish that it turns out well. Right now there is urgent work at hand. If I ever needed to struggle to earn my peace and happiness, now is the time. Men like Zuokumor must be stopped from the path of destruction that they are plotting for our people in league with those who do not wish us well. Since he has declared war against me, I have no choice but to carry the same back to him in his stronghold. I am making a representation to the same Council of Beniotu Chiefs from which he claims to have received his dubious mandate. He is no more Beniotu than I am. Besides, we come from different subclans. He is Oproza, and I am Mein: he may have his mandate from Oproza, but not Mein. You will be hearing from me.

Yours sincerely,
Tonwe Brisibe.

Brisibe Compound
Seiama.

9 February, 1978

Dear Piriye:

Forgive me if this letter is a little rushed. Priboye will be coming here shortly to collect my mail to be delivered to you. I must set down this report before I go out to inspect my fishing lines and traps. Something happened yesterday, after I had written the accompanying despatch, which I think you ought to know at once.

Your friend, Mr Ebika Harrison (Bickerbug, as you call him), visited me yesterday. I wonder why you failed to warn me of his visit. In your last letter, you reported him as saying he had been this way lately and spoken to some people. But there was nothing about him returning so soon afterwards. Besides, although he seemed (from your report) to commend my contributions at the Warri meeting, I did not think I was the sort of person he would want to talk with.

What particularly disturbed me about his visit was not so much that he came. I assumed he knew you and I were close, and felt I would welcome his reference to you as a mutual friend. But it was the sort of company he came with. He had with him some fellows of questionable character, the kind you and I would hardly want to be seen with. Some of them are known to be the most notorious dropouts in this area, who have criminal records or are always in some kind of trouble with the law.

There were about seven men in his company. I recognised Robinson Esiama of this very village; he has been to jail no less than thrice. Then there was Finiba Bozumo, also of this village: they call him "Seadog", because he is known to be constantly going out to sea and robbing ships and even the crew of oil tankers. Also in the group was a young man from Patani by the name of Samson Ekiyo. He is an ace swimmer (they say he can stay as long as ten minutes underwater in a single dive), and has won numerous prizes in various sports festivals in and outside this country. But last year he was dismissed from his job with the State Sports Commission in Benin over some scandal having to do with cocaine. You should have seen his red eyes yesterday. Both he and Bozumo were recently arrested for complicity in a riot at one of the oil rigs around here (some say they actually participated in the riot). But the court discharged them for want of conclusive evidence.

I cannot remember the names of the other men in Harrison's group. But from the way he introduced them to me (one was *formerly* a fitter in one oil company, another was *formerly* a driller, another *formerly* a dockworker, etc.), and from the very looks of them, I could tell they were dropouts, men of dubious record.

My wife was terrified by their entry into our house. I myself was so disoriented I forgot to offer the traditional hospitality, to which I have become thoroughly accustomed. Harrison had on a black T-shirt and dark grey trousers. I recalled your mentioning in your last letter that he wore those things when you visited him at his place. I wondered if he ever took them off. He also had on the dark shades you talked about, and in addition a black beret. The others were not better dressed. When they entered, my wife thought I was in some kind of trouble. She stood close to me where I sat, her arms folded, her face drawn with anxiety.

Harrison first introduced himself, mentioning that you and he saw each other often in Lagos. Then he introduced his men. I told them they were welcome. But I was so eager to terminate their presence in my house I asked them, as nicely as I could, to what I owed their visit. Harrison said they were here on a purely social call. I was a respected figure in the community, one he had always respected ever since he knew me through my work on the *Chronicle*. He said that although he was sorry I had to leave the paper the way I did, he was glad that men like me were back in the homeland to give our people the benefit of our wisdom. I told him I had come home simply to rest. He said he appreciated this, but was sure I could be of service if the need arose. I asked him what kind of service he had in mind. He said he was happy, for instance, at the way I stood up to the enemies of our people at the meeting in Warri. He suggested it was time we made a bolder and more decisive strike in defence of our land and our livelihood. Without bothering to ask him what exactly he meant, I told him that my contributions at the Warri meeting were the best way I knew how to fight; there was nothing that could not be solved by constructive negotiation; there were bound to be obstacles along the way, but these would be overcome if men exercised due patience and discretion. I told him that any other method, such as physical confrontation, was bound to worsen the problem and might do more harm than good to our people; anybody who resorted to violence should expect to face the force of the law, and should expect no sympathy from us.

All along, there had been a faint smile playing over Harrison's bushy face. But as soon as I made that last point, his face fell. In fact I observed certain uneasy movements around his mouth, like twitching. He stared at me for a while. Some of his men sighed, others shifted uncomfortably in their seats. Without further delay, Harrison rose from his seat, bid me good day and good luck, and walked out of my house. His friends rose quickly and stormed out after him, some of them throwing me a menacing look as they passed.

I felt relieved by their departure. In all honesty, it was not without trepidation that I spoke to them the way I did. But it was clear to me that the alternative was complicity in schemes of doubtful merit. I wanted no part in them.

I got up to watch them as they walked away from my premises. They were heading towards the other end of the village. Have you ever seen any of those American cowboy films, in which a handful of bandits hold a whole town to ransom and walk down the main street of the town for a final showdown with the sheriff and his men? That was how Harrison and his friends looked, even without guns at their hips. As she stood nervously behind me at the door, my wife suggested I make a report at the local police post. For a while I contemplated taking that step, but decided to let well alone for now. For the moment, I was engrossed in a little reflection.

I see no good in the ways of men like Harrison. But it is this country that breeds the likes of him. By our reckless and corrupt disregard for the welfare of the people, we are encouraging the growth of forces among us that will destroy the prospect of that peace and happiness that we so earnestly crave.

Excuse me, I must go and inspect my fish traps. I will report any further developments. But be very careful how you deal with Harrison. I do not know what he is up to, but I think he is a dangerous man.

Yours sincerely,
Tonwe Brisibe.

Lagos, 17.2.78

Dear Tonwe,

I was just as shocked to hear of Bickerbug's visit to you as you were to see him at your place. He never told me he was going to the Delta, let alone to call on you. For about a week I checked on him at the two places I had any chance of finding him – at his Obalende residence, which, to keep up the decoy game he's been playing with the NSS, he always leaves unlocked, and then at his Ajegunle hideout, which took me time (and courage!) to locate. His neighbours at the Obalende place couldn't help me, apparently because he has become much less friendly towards them than he may have been in the past. The Ajegunle place is more like a den than a residence. It's a bungalow with two rows of rooms, about three rooms on each side of the central corridor. I'd been there about four times and only found one man once emerging from one of the rooms – he was surprised to see me, almost resentful in fact, and he pretty much denied knowledge of Bickerbug's whereabouts when I put the question to him. Even after I got your two letters from Priboye about six days ago it took me three more visits to the Ajegunle place before I could find him. But find him I certainly did, and must confess I find the atmosphere around him increasingly charged and insidious.

Two days ago, to be exact – that's when I finally tracked him down there, at about 5.45 p.m. I was barely 50 yards from the house when I saw him shaking hands with a stocky, blondish white man who then got into a blue Nissan Patrol and drove away – he passed close enough for me to see his face clearly, but he didn't look like anyone I'd ever set eyes on before. Bickerbug must have seen me coming, but he turned and walked into the house without waiting to receive me at the door.

His room was to the right, the last door at the end of the corridor – I knew because it was the only door open and I simply walked up to it. I didn't have to knock, for the door was wide open. There was a pale, threadbare curtain which did a poor job of shielding the room from public view, and I simply parted it and walked in.

'Hello, Ebika,' I greeted.

He seemed to have been putting something away in a pile of junk at a corner of the room.

'Hey, hey, hey, my good good friend and brother, Piriye,' he turned round to respond, with a most exuberant cheerfulness. 'How are you today? It's been a fine day, hasn't it? But you don't look so happy. What's the matter, man?'

'Nothing,' I said, marvelling at his ebullience. 'Do I look sad?'

'Well, not exactly,' he said, 'but we must learn to wear a happy face all the time. The world can be a sweet place to live in, you know, if you can get yourself together and take care of things. Come on, my friend, sit yourself down and relax. Reee-laaax, over here.'

He threw his arm around my shoulder and led me to the only piece of furniture in the room, two long wooden benches pushed together against the wall to form a sort of bed. I took my seat, and he settled down near me, beaming broadly and exuding considerable warmth. He even slapped my back, and that shook me up quite a bit as I was somewhat disoriented, trying to fathom the cause of his high spirits. I gave him a good look from the head downwards, and he burst out laughing, obviously amused at my disorientation.

'Cheer up, my friend,' he said, rather loudly for my temperament at this point, 'experience the sweet cool air of the harmattan season.'

This elaborate cheerfulness got me wondering if, as on the last occasion we met at his Obalende residence, he was putting on an act for the benefit of the NSS. So, I instinctively bent to look under the benches and see if there was a bug stuck somewhere beneath them. But he pulled me up with a hand on my shoulder and yet another laugh which got me even more disoriented than before.

'Come on,' he said, still laughing. 'Have no fear about this place. Our friends dare not come here. Besides, I have combed this place thoroughly with my own special kind of gadget. I call it the mine-sweeper. Do you know what a mine-sweeper looks like?

I shook my head.

'Well, well, well, how could you?' he said. 'When we were out there fighting the war, some of you fellows were savouring the sweet life here in Lagos.'

'You didn't see that much more of the war than I did. Ebika,' I cut him short.

'Are you joking, me and Isaac Boro –'

'Yes, but you joined Boro only midstream, and after he got killed you came over to Lagos and stayed here for the rest of the war like most of us. Don't forget you used to come into the *Chronicle* to hand us your releases denouncing the government, and –'

'All right, all right,' he said, giving in. 'But at least I saw *some* action. You can't deny me that.'

'Not at all,' I said, and changing the subject, 'Who was that white man I saw you with a moment ago?'

'One of my friends, man,' he said, rising up from the bench and pacing about, 'or am I not supposed to have any friends?'

'I didn't say that.'

'Okay. He's one of my friends. Just ... one of my friends.'

'Does he work here in Lagos?' I asked.

'I *don't* know where he works,' he flared. 'What do I care where he works? I just happened to meet him somewhere and ... he's ... he's just one of my friends, that's all I can say. Just one of my friends.'

'Okay,' I said, calmly now. 'Okay. I get your point.'

For a while we didn't speak a word to each other. I simply watched him as he paced about, his hands now in the side pockets, now in the back pockets of his blue jeans, the next minute his arms clasped about his chest. He still had the dark glasses and the black T-shirt on. Like you, I wonder if he ever took those things off – the only change was from the dark grey trousers to the blue jeans. I think he suddenly realised how unpleasant he was being, so he stopped to beam at me and settled down again near me.

'Come on, man,' he said, putting his hand again on my shoulder. 'Let's not fight over nothing. Now in fact is the time to rejoice and be cheerful.'

'Over what?' I asked, a little perplexed.

'Over what? Over a whole lot of things. Over life, for instance. Over freedom. Over ... over being Beniotu. Over everything.'

It had become a little dark. Bickerbug got up and looked around for a match, struck it and lit up a hurricane lamp sitting amongst the pile of things on the floor. Then he began pacing about.

'What about being Beniotu?' I asked him.

'Like ... for instance, what do you know about our local

Beniotu traditions – our folksongs, our folktales, and so on? What do you know about them?'

'Well,' I hedged, 'it's been some time since ...'

'There you are,' he said. 'You call yourself a Beniotu man, and you can't recite any Beniotu tales or songs. Like ... er, let me see.' He stopped to reflect, and didn't seem to care that I sighed and yawned and showed little interest in the subject. 'Like, what about that fishing song about the mad sawfish, the one that goes *Ekpidi kon bo ya abere, Edaa ekpidi kon bo ya abere, Oki yo lolo indi yo* – the one that plays on the contrast between *ekpidi* for catfish and *ekpeti* for box. Do you know that one?'

I shook my head vaguely, not being too sure.

'Or the other fishing song about various kinds of fish, the catfish, the tilapia, the *aba* – the one that begins *Aba duone opolo a sougha* – you remember that one?'

'Not really,' I said.

'Man, where did you grow up?' he jeered.

'Mostly in Warri. But I'm not that ignorant of our traditions.'

'Look, we *are* a fishing people. You must know some fishing songs. What about that very popular one which the fishermen sing while coming home with a plentiful catch, and the evening tides start up, and their boats are wafted ashore by the bountiful waves, and they celebrate both the fishes and the waves as one large Godsent harvest, and ...'

'The one that starts *Yei yo, Yei yo* ...?'

'Good!' he cheered. 'We haven't quite lost you yet. Now what about some of those moonlight songs, especially those sung by women? Man, I used to find some of them really touching, especially when my cousins sang them. You know how very much our traditional Beniotu girls love being carried away into marriage, how they love for their men to elope with them, especially when the girl's parents don't give their consent to the marriage, but even sometimes when the parents don't really object to it. They think the sweetness of marriage comes most with elopement – for them it's the most heroic way to get married to their men. You remember?'

I nodded just to agree, though I didn't have the foggiest idea what he was talking about. And I didn't really understand why all this stuff excited him so much.

'Now, there's this song by a young girl waiting for her loved one Alowei by the edge of the creeks, where they had planned to meet for the escape. But he was nowhere to be found:

Alowei yo bo bo
Alowei yo bo bo
Tamu toru aruoama
Toru pagala, toru bani ...'

I was truly bored, so I ignored him while he went on singing and dancing. A book with a flaming red and yellow cover, half-stuck into a polythene bag among the jumble of stuff by the corner, caught my interest. I bent over and picked it up to look at it. I had only read part of the cover title – *Blowout! A Handbook on* ... – when Bickerbug broke off singing and snatched the book rather violently from my hands.

'Give me that book!' he snapped, and tucked the book back into the bag, well out of view.

I couldn't disguise my shock.

'What's the matter with you, Ebika?' I said. 'You've become too edgy for me lately.'

'I don't like people prying into my life. A man is entitled to some privacy, you know.'

He was struggling so nervously to zip up the bag that his dark glasses fell off his eyes onto the floor. He hurriedly put them back on and carried on with the zipping.

'Prying!' I said. 'Since when have you started hiding things from me?'

He had finished zipping the bag. He got up and paced up to the door, and stood there with his back to me, his hands in his back pockets.

'So all that talk about being a friend and a brother, all that Beniotu brotherhood stuff – that was just a put-on, wasn't it?'

'Well,' he said, 'I'm just not sure any longer what side you're on.'

'What's that supposed to mean?'

'Man, the company you keep these days – the Ekundayo Bensons, the ... the ... the ...'

'My girl Lati. Why don't you say it?'

'I didn't say that!'

'No, but that's what you really meant to say. So why don't you say it?'

'I *didn't* say that!' he bawled. 'Leave her out of this. I never spoke her name.'

'Well, what do you mean then? You don't really believe I'm Beniotu enough for you, isn't it? That's really why you're trying to teach me all these songs and customs. You don't trust me any more as a Beniotu man because I now keep the company of Ibile people. As far as you're concerned, I've betrayed the Beniotu by getting close to the Ibile. So now you hide things from me.'

'You've got it wrong, Piriye,' he said. 'Think again, man. Think hard.'

'Think what? Was that why you went away to the Delta and never told me a word about it?'

'Must I tell you about every place I go? Am I not free to make my own plans and my own movements?'

I looked hard at him, and I swear, Tonwe, for a moment I felt like telling him the limits of his freedom. But I'm glad I kept myself sufficiently under control over the matter. I simply nodded my head at him and said:

'Is that right? So free you were going to call on my own man down there and you never let me know beforehand?'

'Yeah, your man,' he sneered. '*Man* indeed! I wish I never met the likes of him. Who the hell did he think he was, treating me and my friends the way he did? What the hell – what did you tell him about me to make him so inhospitable?'

'I didn't have to tell him anything, Ebika. He simply sized you up with the pack you were in, and knew he had to be careful how he carried on with you lot.'

'Is he better than the rest of us, the people you call a pack? Such a damn conservative asshole. The Beniotu people don't need the likes of him – they're a drag on our cause.'

'Cause – what cause?' At this point I decided I'd had enough and rose from the bench. 'I'm not going to sit here while you talk like that about people I respect. Yes, he's better than the rest of you and a million more like you who preach all that bullshit about a Beniotu cause. Get out of my way, man, you stink.'

I brushed against him as I walked out of the room.

'I stink! *I* stink? Look who's talking! It's you who stink, Piriye, let me tell you, if you don't know. It's you who stink like shit. You're lost, my friend. When you get home, think hard over what I'm telling you. You're lost, and I wish you had sense enough to get out of the mess you've got yourself into. You're lost, man, lost! Far, far gone astray! I'm sorry for you! Lost ....'

He was still ranting in that raucous, raspy voice of his when I walked out onto the narrow road of the close and headed home. I was amazed to find how dark it was – there really isn't any electricity in that part of Lagos, can you believe that? but I was too upset to feel anything like fear. I simply walked on to a point where I found a taxi. I should have told you – though I didn't really need to – that I'd left my car behind at home because I didn't think it would be safe in Ajegunle, at least that part of it where Bickerbug had taken a hideout. So I got into a taxi and headed straight home. I was so upset I decided not to go over to Lati's place, nor did I even have a proper meal. I stuffed part of a loaf of bread into my mouth, chased it down with a bottle of beer and went straight to bed.

The experience of the evening had made me very sad indeed, and for a long time that night I couldn't sleep. For one thing, I had begun to get out of the habit of losing my temper. In my very delicate relationship with Bickerbug I had promised myself I'd take very special care to keep the atmosphere on an even keel, especially since our previous meeting when he'd reacted so violently to my attempt to pick up his little packet for him. I stand to lose a lot more than to gain by falling out with Bickerbug – not simply in terms of my charge to keep an eye on him, which in spite of what I've seen so far I still don't feel really committed to, but more in terms of our investigative project. I could see a valuable opportunity slipping out of my hands. One of the first practical lessons I learnt from you many years ago was to stay always on the good side of my subject. In fairness to myself, I did try hard to endear myself to him, and even he acknowledged I had stuck my neck out for him. To see this marvellous advantage almost dissolve between my fingers really did fill me with a sense of failure.

I was also quite worried when I came to think a little more of some of the things he said, and the way he's been acting towards me lately. Could it be that he had discovered I'd been charged by the NSS to cover him? A man who could set up a communication line from a detention camp to the world outside, even to the enclaves of policy, a man who could get hold of the transcript of a meeting far away in Warri when he had barely got out of detention, a man capable of so much undercover activity – I'd be a fool to think he'd find it hard to sniff out the moves that the NSS and I have been making around him. The way he looks at me rather penetratively from time to time, it's perhaps only those dark shades of his that have saved me from seeing the accusatory gleam in his eyes. And even though he would seem to have kept me in his confidence by letting me know the places where I could find him – whether out of regard for our long acquaintanceship or for a lingering thread of ethnic solidarity, I'm not really sure – it began to dawn on me that he was being very careful that I didn't get close enough to his privacy to do him any real harm. Who knows – maybe the antics he has been displaying lately, especially all that song and dance act of our latest meeting, were simply his way of fighting the embarrassment which my bad faith had created between us. I began to hate Yelwa and the NSS more and more for putting me in such bad humour with people who meant a lot more to me than they did.

Things got worse the next morning, that was yesterday. I was woken up about 6 a.m. by Lati, who was worried that I hadn't come to her place at all the day before, against my usual habit. I'd gone to the door to answer her knock and let her in. As I was making my apologies about not looking her up, my eyes fell casually on the dining table. Let me tell you a bit about that table. In the years of the deteriorating relationship between my ex-wife and me, I would rustle up a quick meal for myself and sit there to eat it, alone. Since she left, and the affair between Lati and me has grown steadily, I have seldom visited the table. Most of the time these days I eat at Lati's place, or grab a snack somewhere. Once in a while I buy a loaf of bread or a packet of biscuits for a quick munch.

So the previous night I'd pulled out that loaf of bread and taken some of it with the beer, but carelessly left the rest of the loaf on the dining table – although I hadn't sat there to eat (I'd done so in my bedroom, where the bottle of beer still was). Now as my eye fell on the table after I'd let Lati in, I saw a cockroach nibbling at the bread, and I dashed to slap it dead with a book. It escaped my blow, and crawled underneath the table. I bent under the table to pursue it and there, in between the supporting woodwork of the table, I saw an object. I tipped the table over and, ignoring the cockroach, concentrated my attention on this object which looked exactly like the bug that Bickerbug had under his bed at Obalende, except that this one was maroon in colour.

'What is it?' asked Lati, who stood behind me with curiosity written all over her face.

I gave her a hush sign, and quietly unstuck the object from the table, setting the table back on its legs and getting up to study the thing. Like Bickerbug's, it had a little on/off switch, which I moved from on to off.

'A bugging device,' I told Lati.

I'd told her recently about the one under Bickerbug's bed, so she didn't need to be told what was the purpose of this one. But you can well imagine the feeling that was building up in me as I held that thing in my hand. So Yelwa and his men had to do this to me too? I tried to figure out how the bug could have been planted in my apartment. I knew of course that the NSS had found their way into this place before. They had taken my car away for repair while I was in detention, and brought it back to me – all with my own keys, which I'd left behind in the apartment when I was being taken away by Osawe and Haastrup about three months ago! Now when I returned from detention one of the first things I did to regain my privacy was to change my front-door lock to something a little more difficult to fiddle with – a digital lock. So, either the bug had been installed while I was away in detention, or the digital lock is not as burglar-proof or break-safe as I'd thought.

Well, I was very upset when I discovered the bug. Just to be safe, I combed as much of the apartment as I could with Lati, searching for bugs. We found no more. Then I had a quick bath while Lati fixed me something to eat – *dodo* with stew, and coffee. While I dressed and then ate, I told her in a nutshell about my encounter with Bickerbug. Then I took off and went straight to the Security offices to see Yelwa.

As I was driving to the place it occurred to me that I had to be careful how I confronted Yelwa. How would I know how much he and his men knew already? Since my release from detention and the pact between me and the NSS, I'd only been there once to report that I hadn't seen Bickerbug but thought he was in town and would soon have something concrete to report on him – was it not odd that they hadn't complained I wasn't showing sufficient commitment to our pact? And although I spent far more time with Lati at her own place than in my apartment, how could I tell how much information the bug had transmitted to Security since it was installed in my place? Would it not be wiser for me to come out with some information, however tailored, than to keep on pretending I didn't know anything or – worse still – come straight on at Yelwa with a show of righteous indignation at finding the bug in my house?

Which was precisely what I did. When I appeared before Yelwa I greeted him with a certain guarded warmth. He had on his usual false cheerfulness, and believe me I did everything I could to avoid losing my patience with him.

'How are we today, my good friend?' he beamed.

'Fine,' I said.

'Is your car okay? No problem with it?

'No, none at all,' I said.

'Good. Now, what do you have for us?'

'Well', I said, 'not much. But I was with Mr Harrison a few days ago. I visited him at his place and he took me out for a drink.'

'Oh, jolly good. Where did he take you?'

I was caught by that question – I hadn't anticipated that line of manoeuvre.

'Well ... somewhere near his place,' I stuttered. 'A little bar not far from where he lives, I can't quite remember the name of it.'

'Okay,' said Yelwa, 'it doesn't matter. Carry on.'

I assumed that with the bug safely out of Bickerbug's room on the night in question, the NSS could not have picked up any information from it beyond the invitation to a drink loudly given to me by Bickerbug before he switched off the device and I took it out to the backyard. So I simply told Yelwa all I wanted to tell him. I said Bickerbug had complained about how badly he had been beaten on his first day in detention; how very bitter the whole experience of the detention had made him; and how he wished he could teach his captors a real lesson. I said I'd urged him to forget the past and try to settle down again; that Bickerbug had declared that even if he didn't cause any trouble for the government, he was nevertheless not going to stop drawing the attention of both the government and people of this country to the appalling situation of oil pollution in the Delta areas. I even told Yelwa about Bickerbug's new interest in civil and petroleum engineering, and how he'd lectured me very absorbingly that night about oil pollution on our way back from the bar! I finished by reporting that Bickerbug had told me he'd be going to his village for a few days to see his relatives and cool off.

You must understand why I told Yelwa about Bickerbug's interest in oil technology and pollution and even about the books. Since his room remains ever unlocked, I felt certain that anybody could have gone there at any time and looked through everything there – so what's the point pretending the NSS don't know anything about these things? I took a chance on not saying anything about Bickerbug's hideout in Ajegunle and about the whiteman I'd seen him with. Perhaps I was being naive in suppressing all that, but I could always claim to be waiting to find our more firmly about this association before making a report on it. At any rate, I was in such a mood with the NSS yesterday that I didn't feel like giving them any more information than I had to.

After I'd finished Yelwa thanked me for my cooperation with them, and said he hoped I would continue to be diligent and report to him at once any suspicious movements on the part of my friend. He then said I should expect something very soon from his office as a token for my efforts. I knew very well what he meant by "something", but I didn't express any alarm at the idea, partly because not having objected to their repair of my car I didn't have much moral basis for objecting to any other form of bribe now (how I hate the word!), but especially because I was at this point trying to summon up the proper frame of mind to lodge my protest. He must have read the effort on my face, which had been staring steadily and cheerlessly at him as he made his exhortation to me, for he stopped speaking abruptly to look at me with some consternation.

'What's the matter, my friend?' he said. 'You don't look very happy.'

Still staring glumly at him I fished the bug out of my jacket and laid it on his table.

'Officer,' I said, 'was this really necessary? Why did you send your boys to bug my house?'

He picked up the thing from the table and turned it over in his hand, studying it with wide-eyed and open-mouthed wonder as though he'd never seen the like before or as if horrified that anyone should do this to me.

'How can you give me an assignment and then tell your boys to trail me? After all, you and I came to an understanding, which I accepted like a gentleman even though I didn't particularly like the idea of spying on my own friend. But I've tried to convince myself that I was doing a national duty. At least you could have given me a chance to show whether of not I would keep my side of the bargain.'

Yelwa dropped the bug on his table and put one hand to his bent head, rubbing and shaking it with a show of disappointment or maybe regret.

'Oh, no, no, no,' he moaned. 'Those foolish boys! Why did they do this, eh? I'm really sorry, Mr Dukumo, Allah! I want to assure you I didn't know anything about this. Where are those boys – Suleiman!'

Not content with the loudness of his voice, he pressed the buzzer on his table. In a short while a policeman opened the door and saluted as he appeared.

'Call Obiozor,' Yelwa commanded, his head still bent.

'He's not in, sir,' said the policeman, at attention.

'Then call Haastrup or Adigun.'

'None of them is in, sir.'

'All right. You can go.'

The policeman saluted and left. Yelwa raised his head and looked me with contrite eyes and clasped hands.

'Leave this to me, Mr Dukumo,' he said in a tone of deep sincerity. 'I'll find out who planted the bug in your place. But please understand that I had nothing to do with it. Somebody has obviously gone beyond his duties. Leave it to me, Allah! I'll make sure this kind of thing does not happen again.'

I sighed and nodded. We both knew we were lying to each other, but it was convenient to keep up the pretence especially in the absence of any incontrovertible proof of bad faith on the part of either of us.

'Okay,' I said, and rose to go.

'I'm very very sorry,' said Yelwa, rising too, and stretching out his hand, which I took. 'It won't happen again, I assure you. *Wallahi!*'

'I believe you,' I said. 'Bye-bye'.

'Bye-bye. Be expecting that thing.'

'Okay.'

You can imagine how confused I felt as I was leaving the Security offices on my way home. Although I had good reason to feel alienated from Bickerbug, given the new development in our relationship, I was not about to turn my back totally on him and embrace the NSS. My discovery of the bug in my place definitely erased any illusion I may have had about cooperating with them in any way. And if it was true as I feared, that Bickerbug had discovered my collusion with the NSS against him, then I could do no better than to find some way of reconciling myself with him and making him realise I never took my relationship with the NSS seriously anyway. But, I also felt, he must come to terms with my new personal circumstances and concede that, while not in any way abdicating my consciousness as a Beniotu man, I had nevertheless found for myself a source of happiness which had forced me to reassess my old feelings about ethnic relationships. I thought he should at least be charitable enough to grant me this.

I told you I was confused, Tonwe, and I really am. Being a paid spy on my friend and tribesman is not a role I enjoy playing, however grudgingly. I'm very sad about that, and I hope in the next day or two to go back to Bickerbug and make my peace with him, especially now I know the NSS is just as much on my tail as they expect me to be on his. But I'd be lying if I told you his recent moves and schemes didn't bother me. To start with, the gang he visited you with says something to me about his present frame of mind, and in spite of myself I almost wish you'd reported him to the local police. Besides, the *Blowout* book he snatched so vehemently from me – what is it a "handbook" on, or was there something stuck between its pages he didn't want me to see? I don't know, I'm truly worried about the present shape of things. I want to go back and make my peace with him because he's a friend and tribesman as well as a valuable professional subject. But I'm also sufficiently anxious about my safety and conscious of my changing circumstances – Lati and I are going to be parents! – not to allow Bickerbug to stem my curiosity about his plans.

Talking about Lati brings me to a matter that's given me more worry since I got your letter – I mean the first of your last two letters – than any of the problems I'm currently battling with. I can excuse a beleaguered man like Bickerbug when he makes snide remarks and insinuations about my present ethnic company, but it hurts me very deeply when a man like you, for whom I have the utmost respect and to whom I've confided everything, has to make apologies to me for your feelings about the Ibile. Frankly, Tonwe, you cannot know how sad it makes me feel to think that anything can drive a wedge between us at this point.

Isn't it strange, though, what's happened to us within the last year and a half that we've been involved in this business? I suppose I should say I'm sorry I roused you from your retirement – in which you were content to be at peace with yourself and with a country that had done you wrong – with a project which has only succeeded in confronting you with an experience and a sentiment you had chosen to put behind you. Of course, I've never had any doubts about the ethnic or nationalistic element in this project since the first time I proposed it – you remember what you said to me then? – but little did I know it would take this turn. In a different atmosphere and a better frame of mine I could conceive a fitting title for a chapter in our book that would do justice to this stubborn ethnic factor. But you've never really wanted to be bothered with these titles anyway, nor can I be sure now if you think the project itself is any longer worth your while. So let me assure you that I'd be willing to let this project rest a while rather than lose the regard I've had for you for about a dozen years.

Lati gives you her warmest wishes. We've fixed the traditional wedding engagement for 5th March in her home town Ilesha, where her parents live. Even if you can't come, we'd be happy just to know you're with us in spirit. Believe me, Tonwe, I've never been so happy in all my life.

Very sincerely yours,
Piriye

Brisibe Compound
Seiama.

24 February, 1978

Dear Piriye:

By now you must have heard the terrible news. No less than five oil installations in the Delta have been destroyed by bomb explosions. These include three offshore rigs like the one at Ebrima near my village where, you will recall my telling you about a year ago, Opene and his companions had been assaulted by men of the naval patrol. The huge storage tanks at Apelebiri near Angiama, Harrison's village, and at Ogbodobiri have also been blown up. You cannot imagine how much oil is floating about now in these creeks. It is better seen than described.

There is no doubt in the mind of anyone here who is responsible for all this. The police and navy mounted a quick search for Harrison and his gang as soon as the explosions took place. It appears Harrison left this area before the event. But some of his men have been rounded up. Robinson Esiama and Seadog Bozumo of this village are under arrest now. They have also got a young engineer who had recently been laid off by Freland Oil. He is said to have been personally responsible for the explosion at the Ogbodobiri storage depot. I think his name is Ernest Tobi: I am not sure if he was one of the party that Harrison brought along to my house. Samson Ekiyo is said to be at large. But his poor old father, a retired ship's steward, has been taken away on the accusation that he was concealing his son's whereabouts. A few other people connected with the culprits in one way or another have also been taken away. There is a general feeling of fear and insecurity in this area. The arrests have become rather indiscriminate. Everybody is wondering who will be the next to be taken away.

My dear Piriye, in the name of God, please go to the Security police and report Harrison to them, if you have not already done so. You can no longer afford to keep on protecting this criminal on the ground that he is your friend and tribesman. I have never believed that the tribe needed violence to justify itself or to press its grievances. At any rate, it is in your interest to turn Harrison over before the authorities find out you are protecting him from them. I made the costly mistake of not reporting him when I had a chance to do so. Please do not make the same mistake, or he may do more serious damage.

I am sorry you took my statement about the Ibile to imply I have reservations about your present affiliations. Nothing can be further from the truth. If God in His infinite wisdom has chosen to bless you now with an Ibile wife, who am I to stand in the way of that wisdom? Let me also assure you that I am just as anxious as you are to pursue this investigative project to its logical conclusion. How can we, who know so much about what has happened, deny that knowledge to a world that may already have convinced itself of our complicity in these events? At any rate, please report Harrison to Security right away. I am not sure we are making it any easier on our consciences by pretending that the labours of investigative journalism are more urgent than the safety of the country.

My wife and I wish you and Lati a happy ceremony at Ilesha. Our hearts go with you.

Yours sincerely,
Tonwe Brisibe.

P.S. Has it occurred to you that the Bernstein and Woodward book, which you seem to have chosen as a model, does not have any chapter titles?

T. B.

Lagos, 26.2.78

Dear Tonwe,

I haven't found it easy summoning up the strength to write this letter. Even now, I am only writing because Priboye tells me it will be at least one month before he comes to Lagos again – the shipment he was awaiting has arrived and he has to be in Warri for some time supervising the construction there of his fish storage depot. But my heart is sore and my body is terribly weak. Lati has been away for several days and has not returned. I fear the worst has happened.

On the 19th February I went to see Bickerbug to try to make peace with him, as I told you in my last letter I would. He was nowhere to be found. I went first to the Ajegunle place. The door was open, but there wasn't a thing in the room – it was all cleared out, not even the two long benches by the window were there. It was about 3 p.m., and there was no one there I could ask any questions about him. I got in my car and drove on to the Obalende place. Same thing – all cleared out, no bed and no books, none of the usual junk anywhere around. None of his neighbours knew where he was or when he moved out. There was no doubt in my mind – he had gone!

When I couldn't find him, I thought he'd lost faith in me and decided he'd no longer tell me his whereabouts. It was only later that day that it began to dawn on me he'd gone not so much out of circulation as underground. I was at Lati's place discussing with her what Bickerbug's disappearance might mean and what was to be my next line of approach, when the NTV news came on at 7 o'clock. It was reported that a section of Lugard Bridge – that old colonial monument – had been blown up during the 4 o'clock rush hour. It's a waste of time describing the horror! If by any chance you've renounced your old resolve to stay away from newspapers – I don't see how in the present state of things you can continue in that resolve – but if you can manage to check the newspapers of the 21st and 22nd you'll get a picture of what terrible damage was done. It's horrible, Tonwe, simply frightening!

I didn't have the courage to get up and go anywhere. To tell you the truth, I was frightened. What would I tell Yelwa if I saw him? I decided to sit back and think. Lati and I spent most of the night worrying if Bickerbug could have been responsible for the destruction.

'I don't see what business he has blowing up bridges in Lagos,' I remember her saying. 'His fight is not in Lagos.'

'Yes, but you remember that long lecture he gave me on bridges and oil rigs,' I said. 'If he didn't have any plans involving bridges why would he spend so much time studying them?'

'Why would the bridges he planned to blow up be in Lagos?'

'The man is a maniac,' I said. 'He's bent on revenge, and the bridges might be just as well in Lagos as anywhere else.'

'I don't know,' said Lati. 'Even a madman has his reasons.'

As I said, we spent long hours of the night agonising over the problem. But one thing I finally decided was that I was going over to Yelwa in the morning to tell him a little more than I'd previously allowed myself to. I had to find a way of doing this without raising questions in his mind or inviting some form of censure – if not worse – from him.

That night I slept at Lati's place. Early the following morning she turned on the radio for the six o'clock news brief. The only thing of consequence that was reported then was the Lugard Bridge incident and the rescue operations that were underway. But about 20 minutes later, during the musical interlude that followed the news, the presenter stopped the music to flash the report about explosions at the oil installations down in the Delta. At this point there was no doubt in my or Lati's mind who was creating all this havoc. The only thing we couldn't agree on was the role of the bridge in the whole scheme of destruction. She thought the bridge explosion was simply a decoy to confuse Security as to his whereabouts. But judging by the fact that news of the explosions in the Delta had only just reached Lagos, I was inclined to believe the incidents in the two areas took place at almost exactly the same time – trust Bickerbug for the cold-blooded neatness – and were of equal importance in his scheme of revenge.

Anyhow, Lati and I left the house about an hour later. I dropped her off at the Iddo end of Lugard Bridge to find her way through the terrible commotion of crowds and policemen and firemen, with cars flashing their rooflights and sounding their sirens, and policemen barking orders to the surging and wailing crowds to clear the scene for rescue operations. I was there for a while trying to see if there was anyone around from Security, and when I didn't find anyone I drove on to their offices. It was about 8.30 when I got there.

Yelwa was almost like a madman when I saw him in his office. He was not wearing any uniform, and his hair was hardly combed. He was barking orders into the telephone and his face had none of the cheerfulness he usually wore when he wanted to put on his act for me. When he had finished issuing orders, he turned to ask me, without the least pretence at courtesy and indeed with some fury:

'My men have been trying to reach you since yesterday, Mr Dukumo. Where have you been?'

'I spent the night at my fiancée's place,' I said, as calmly as I could, to avoid anything like a confrontation. 'Has anyone gone to get Harrison?'

'No,' he said. 'I sent two men to his residence last night and they reported he had vacated the place.'

'Vacated?' I asked in mock disbelief. 'Okay. I think I know where else we can find him. Give me one or two men.'

'Why didn't you tell me this before?' asked Yelwa.

'He only told me two days ago,' I lied. 'But he wasn't there when I called on him later in the day.'

He pressed the buzzer on his table. A few seconds later Suleiman appeared and saluted.

'Which of the SOs is around?' Yelwa asked him.

'Ogbuibe and Haastrup are in their rooms, sir.'

'Call them in.'

Yes, sir.'

Suleiman saluted and left. *God*, I thought, *Haastrup*! About two minutes later, Haastrup came in with his colleague, whom I'd never met before. Haastrup and I eyed each other with anything but mutual admiration, and I nodded at the other officer with guarded courtesy.

'Mr Dukumo will accompany you two to a possible hideout of the suspect,' said Yelwa. 'Go on, gentlemen, and make sure you bring him here – alive.'

I drove them in my car, with Haastrup in the back seat, of course. What Yelwa had said about bringing Bickerbug back alive made me quite uneasy, because it must mean my two companions were carrying guns although I didn't see any evidence of these. Nonetheless I hated the presence of Haastrup in my car, and I just couldn't help letting him know it.

'They had to bring *you* along, didn't they?' I turned briefly as I drove to ask.

'Look, Piriye,' he replied, 'you've got your job to do and I've got mine. We don't have to talk to each other, you know.'

'But why you?' I persisted. 'I suppose it's always you when the filthy jobs have to be done.'

'I didn't hear you raise any objection when the DSP picked me. So stop bothering me and keep your eyes on the road.'

'I have no control over Security's decisions and you know that. But if I had my way you wouldn't be on this mission.'

'Thank you very much,' he said. 'But remember that we wouldn't be in this mess today if it wasn't for you. If they had let us handle this problem the best way we know how, we could have stopped your lunatic friend from creating all this havoc. But what did you do, Mr Freelance Journalist? You talked your fine grammar with him and looked the other way while he did this thing.'

That stung me, and I swung round at once with a vulgar word ready at my lips. But Ogbuibe intervened just in time.

'Will you two stop this nonsense and let us get on with our assignment? Mr Dukumo, please do me a favour and concentrate on the road. I have no wish to die for this or any other "noble' cause. I have a wife and two children at home who need me alive, okay?'

I accepted the truce grudgingly and drove on. By this time we were at the Marina end of Mainland Bridge. Movement was slow because all other traffic in and out of the Island had been diverted here since the Lugard Bridge disaster. It took another two hours before we finally made it to Ajegunle. When we got to the close I advised that we should park the car at the turn-off into it and walk to the house where Bickerbug had stayed, but the others insisted we should drive right up to the doorway of the house and storm an entry. I didn't like the style – besides, the close was a rugged road and I couldn't be sure that in their present disposition towards me Security would agree to repair any damage to my car. But I didn't want any more trouble at this point. So I drove the car as best I could to the doorway of the house.

Coming up here was a hoax, as you must have guessed – simply intended to give Security the impression that I was taking my assignment seriously. Otherwise how could I have explained the ease with which my "friend" had slipped through my fingers? If my companions and I went up to Bickerbug's room and found it empty as I'd done the previous day, at least there'd be reason enough for them to feel he'd left the place to escape being arrested for his crime. So we charged into the house, with me leading the way, and headed straight for the room. To my utter surprise – or should I say dismay – I found the room now fully occupied. A middle-aged man was lying on a bed, with a wrapper around his waist and his body bare otherwise, snoring away. It was almost noon. I shook him by the shoulder, and when he started up and stared at me, I hardly knew what to say.

'Where's the occupant of this room?' I asked him, with more disorientation than menace on my face.

'Eh?' he returned, startled and confused. He sat up properly as he discovered I was there with two other men who looked even grimmer than I did.

'I said, where's the occupant of this room?' I repeated.

'Oga, I no know which kind ... I no be ... wetin you say make I do?' he was stuttering, becoming increasingly frightened now, fumbling awkwardly.

'I say wey the man wey de stay for here?' I asked again.

'Oga make you no vex,' he pleaded. 'Na me only one with my wife and pickin de stay for here. My pickin go school, my wife dey for ... Adesua!' he shouted.

'Wetin?' his wife answered from the backyard. She was obviously cooking, as we could tell from the smoke that wafted in through the back door. She entered in an instant, and as soon as she saw us she froze. 'Na wetin?' she asked again, eyes wide open with fright, as she wiped her hands on her breast-high wrapper.

'I de ask your man say wey the person wey de stay for this room?' I repeated to her.

She was less disorganised.

'Na only we two dey here with our pickin. No other person de live with us. Na today today we just pack come begin stay for here. I beg make una no vex. Na wetin de happen?'

'Wey the man wey been dey here before?' I said.

'We no meet anybody for here-oh,' said the man now. 'Na true I de talk. If I lie make God punish me.'

I was helpless and confused, because I knew they were right

but I hated them for making me look so stupid.

'This is incredible,' I said, turning to Haastrup and Ogbuibe. 'I don't believe this room could have changed hands within only a couple of days, just like that. At least they must have come to negotiate for the room, and if they did they must have seen the man who was staying here before.'

'You're sure this is the room?' Ogbuibe asked me with a cold stare.

'Absolutely sure,' I said. 'I visited this room before.'

'All right, no problem,' Haastrup said, taking over the interrogation. 'Oga, you say you no know who dey here before?'

'No, sir,' the tenant replied. 'I swear I no know.'

'Wey the man wey rent the room give you?'

The man looked confused and seemed to be seeking for a subterfuge. He stuttered something inaudible.

'Oya, make you follow we come for station,' said Haastrup, seizing the man by the arm. 'If your wife wan' see you, tell am make 'e bring the landlord wey rent the room give una.'

The man worked his arm free from Haastrup's grip, fell on his knees, and started pleading.

'I take God beg you, oga, 'e never tay wey we from Benin come. We no know anybody for here, we no do anything. I no get better work self. I beg you make you ...'

'Shut up and get up!' Haastrup barked out, and gave the man a resounding slap which sent him sprawling to the floor. 'Get up, I say!'

The man tottered up, struggling to avoid Haastrup's wrath and particularly his grip. As soon as he was up, Haastrup grabbed him by the waistline of his wrapper.

'Come on, let's go!' he said, dragging the man away. The man began to wail and plead, and his wife joined him in his lamentation, crying uncontrollably and begging Ogbuibe and me to prevail on our companion and have mercy on them. Ogbuibe was already on his way out of the room. I could no longer stand this heartless show of power nor suppress the growing sense of guilt in me. I held Haastrup by the shoulder, and gave him a look that was intended to conjure pity in him.

'Wait a minute, Haastrup,' I said, 'Take it easy with ...'

'Let go of me, will you!' he barked at me. 'Look, I told you you had your job to do and I had mine. Your job was simply to bring us here. Now you've done it, so get out of my way while I do mine. You've given us enough trouble already, now lay off!'

'Come on, man,' I said, 'try to be reasonable. These people ...'

'I said, lay off! You had your chance. Now let me do my job my way.'

'Okay then,' I said. 'You can take him to your station yourself. I'll have no hand in this.'

'Oh yes you will,' he said.

In a flash he pulled out a revolver from the breast pocket of his jacket, cocked it with the other hand, and pointed it at my head. He had to release his prisoner to be able to do this, but the poor man was so frightened he was rooted to the spot on which he stood. Haastrup's hands were shaking as they held the gun to my head.

'Now move!' he ordered me. 'Get in the car and drive us to the station.'

I kept absolutely cool as I stared coldly now at him, now at the gun. Slowly I nodded my assent, took my keys from my pocket, and walked towards the car quite calmly. From behind me I heard Haastrup give his prisoner a hard push. The poor man escaped falling by grabbing my back. He started wailing again, and so did his wife. As soon as we were all in the car – Ogbuibe with me in front, Haastrup and his prisoner behind – and I started the car, the wife became even more uncontrollable than before, crying in a combination of Bini and pidgin and appealing to their neighbours on the close for help. I could hardly reconcile myself to my complicity in this show of shame.

At the Security headquarters, Haastrup lost no time in turning the man over to detention. I simply had to walk away from the scene to avoid the shaft of guilt which every note of the man's wailing protestations stabbed into me. As soon as I was able to gain audience with Yelwa I made it clear to him I'd have nothing more to do with this case unless I was allowed to locate Bickerbug myself, and especially not with the likes of Dayo Haastrup in attendance.

'And how will you bring him to us if you find him?' he asked. 'We cannot allow you to carry a gun even if you know how to use it.'

That made sense, and set me thinking fast.

'Give me a bug,' I said. 'If I find him I'll engage him in a conversation that will give you the clues you'll need to round him up.'

He thought for a while, and nodded his consent. I got the bug from him, thanked him and left. I was in no hurry to go anywhere but home. I needed to lie down, think the whole matter over and take some hard decisions. Despite my agitation and my hunger, I slept for a very long time that afternoon.

At about five o'clock I got up, washed my face, downed another piece of bread and some Coke, and drove out. The bug was in my jacket's side pocket, but where would I go? Bickerbug was not known to visit many people, not even tribesmen or relatives from the same Beniotu homeland for which he claimed to be fighting. Everyone knew he was in trouble, but even if he could be located or traced through any one of them, how would I put the question without being greeted with a suspicious look? I could of course visit a few Beniotu families in Lagos – Dakoru, Egbuson, Ebiegberi, Porbeni – and conduct the whole discussion in our language without mentioning any names. They would deny any knowledge of Bickerbug's whereabouts, and they might be right and honest. The whole discussion would be transmitted through the bug to a recorder at the Security offices. But suppose I was asked later to identify these people despite their denials? Would I have the courage to resist the demand of officers who were growing increasingly distrustful of my role? And if I gave them any names, how could I stand the outrage of my people when sadists like Haastrup hauled one Beniotu patriarch after another from their peaceful homesteads to detention?

I decided I wouldn't lend myself to such shameful service. I drove aimlessly through mainland Lagos for about two hours. Once in a while I would park my car to mix with crowds at busy points like Tejuoso, Oshodi, Ayilara and Yaba bus stop just to get some sound into the bug and give the impression I was working hard. Of course I was determined now to nail Bickerbug if I spied him anywhere. The maniac had done enough harm already and would get a whole lot of innocent people into more serious trouble if he wasn't stopped. But I wasn't going to provide the evidence that would bring about that trouble.

Lati came to join me at my place about 7 p.m. It was such a relief to see a friendly face again. After the initial moment of courtesies and guarded tenderness she set about trying to fix us something to eat. I suggested she needed to rest for a while considering she'd had a busy day and she was pregnant besides. She simply laughed at my anxieties over her and gave me a reassuring kiss on the cheek.

'Mr P, I'm a working girl,' she said. 'And being pregnant doesn't make me an invalid.'

After supper we listened to some news on the TV, then retired for the night. I'd since switched off the bug, but just to be doubly safe I covered it up in an empty porcelain cup and laid the whole thing away on the far end of a bookshelf in the living room, then shut the door of the bedroom.

I told you Lati and I stayed up most of the night before, didn't I? Well, on this particular night we practically didn't sleep at all. That was the first time Lati and I had a real argument – I mean, a serious disagreement. Let me cut a long story short, Tonwe, for I'm getting quite tired now. The issue at hand was whether I should take my job as a journalist more seriously than my commission as a security agent. For now that it seemed clear that Bickerbug had begun to put his scheme of destruction into practice, I had to make up my mind whether I should work more actively to stop him on his evil mission and deliver him up to the authorities, or face up to the assignment I'd set myself in the first place, which was to justify myself professionally by producing a gem of investigative journalism the like of which had never before been seen in this country. I argued that at this point I'd committed myself so deeply to the authorities that backing out now would be not only dishonourable but indeed dangerous, considering the sort of characters I was dealing with in the security organisation. She countered by asking how much of my commitment that organisation deserved anyway, when it had demonstrated all too clearly by bugging my own residence that I was just as suspect to them as the man they had set me to tail – was it not more profitable for me to pursue professional fulfilment than to get more deeply involved with a system to which I was clearly an outsider?

'What use is fulfilment if I'm dead?' I asked.

'I know, Mr P,' she said. 'But what makes you think you are safer in than out?'

'At least they owe me protection for as long as they know I'm doing a job for them.'

'Okay. But do people like Haastrup know that?' she said, sitting up now. 'What is there to stop him from pulling his trigger on you one day – God forbid! – and telling his bosses the shot was actually directed at Bickerbug but that you were trying to protect him?'

The possibility filled me with horror, more so because it was so distinct. But I was just as baffled by the problem of where Bickerbug might be. Even if I was to give more attention now to my calling than to my commission, at least I had to find my quarry first.

'Let's face it, Mr P,' Lati said, lying down again beside me on the bed, 'I don't think your friend is still in this town.'

'Why do you say that?' I asked, turning to her with some bewilderment.

'I'd be surprised if he was still around,' she said. 'I think he's too smart to stay here. I told you the destruction of the bridge was a decoy, didn't I?'

'Yes, you did.' I was still looking at her.

'Hm. I suspect he's working on his next target right now.'

'Go on, I'm listening,' I said, still gaping at her, my hands interlocked under my head on the pillow.

'Let's see,' she said. 'On the night he gave you that lecture, he talked about bridges and oil installations, right?'

'Right.'

'What else did he talk about?'

I thought for a while.

'Well, nothing really. He talked about the flushing of the tankers, or ballasting as he called it. Then he talked about ... he talked about' – at this point I was forced upright by the growing revelation – 'he talked about the Kwarafa Dam. God, Lati, you don't mean ...'

I was scared to spell it out.

'I'm afraid so,' she said, almost inaudibly, eyes staring blankly.

Oh, God! I thought, and slumped back on the bed, feeling weak. Several ideas were clashing furiously in my head, propelled more by terror than any other sensation. How could I have missed the connection between the various things Bickerbug had talked about, when they were the original agenda of the Committee formed by him and Siekpe and the others? But what on earth could I do about it now, or how would I get it across to Yelwa and his men?

'Well, that's too bad,' I said.

'What do you mean?' asked Lati.

'I can't go any further than this. It's too much for me.'

'Are you giving up your story, then?'

'Well, what am I supposed to do?' I said, sitting up again. 'Follow a madman as he blows up a dam? And what would I tell the NSS I was doing there with him, assuming I came out of it alive?'

'I'm not saying you should go there by yourself. You could ...' she stopped briefly to think. 'Let's see.' She too sat up now. 'Suppose you do this. Suppose you tell Yelwa some more about the lecture your friend gave you that night – tell him the details had escaped you but started to return to you as events unfolded themselves. Tell him about the tanker and about the dam. Request that armed police be posted to protect tankers, while another team is sent to the Kwarafa Dam. I don't think Harrison will have any more to do with the Delta – he has taken care of that sector and must know there will be some security building up there now. I have a strong feeling he'll strike the dam next. You can ask to accompany the armed team to the dam, as he may be on his way there now if not there already. You will be achieving two things – doing your job for the NSS, and grabbing the scoop.'

'What scoop, Lati?' I almost screamed, getting up from the bed and swinging round now to look her fully in the eye. 'Are you serious? Suppose he blows up the dam in our faces just as we arrive on the scene?'

'Mr P, you don't have to go right up to it with the NSS men. You can maintain a safe distance – they are armed men, you're not.'

I shook my head violently as I paced about.

'No, Lati,' I said. 'No way. I'm sorry. I can't do it. I won't risk my life for any of that. Not me.'

That's how we went on, for the rest of the night. There were moments when I raised my voice, but apologised and cooled down again. And still the problem agitated us. Even when we didn't speak we sat up, or lay down with our eyes staring blankly up at the ceiling. At one point I drifted off into a nap, but soon after I woke up with a jump and saw Lati pacing about with a cup of cocoa in her hand. I asked her what the matter was, and she said she couldn't sleep and was trying to calm herself down.

We spoke very little the following morning, we were so tense. In all honesty, Tonwe, I was worried about how best to broach the matter with Yelwa in the next few hours but I was even more disturbed about what the whole business was doing to the harmony between me and my woman. I had never had occasion to argue with her so vehemently before, certainly not to the point of our losing sleep. I was happy when the morning came, because it brought smiles to our faces when we realised we didn't have to do this to each other. We had a bath together and ate the breakfast she made. We hugged and kissed and made up. I drove her to her house to have a proper rest for the next few hours while I drove over to my mechanic to have my car fully serviced and plan my approach to Yelwa. As I said, we made up, but I could tell within me it was an uneasy truce.

It was about noon when I got to the Security offices. I asked to see Yelwa, but was told he had just gone into a top level security meeting with the Inspector-General and other security chiefs. Knowing how long these government meetings usually last – with all the attendant comforts of beverages and meal-size snacks, no matter how urgent the matter might be – I decided to go away and plan my mission a little more constructively. I drove back to Lati's place. She wasn't there, but her aunt, Mrs Kumolu-Davies, was in. She had gone to work, Auntie said, but had left me a note. My heart skipped a beat – Lati had never left me a note before. I practically tore the note from Auntie's hand. It read: *Mr P, my love: I can understand why you want to avoid going to Kwarafa. I think you are right. But I know how much the story means to you, and I'll do my best to get it. Don't worry about me. I'll be careful. Haven't I always been? (Smile!) Will be back as soon as I've grabbed your scoop. Be careful yourself. I love you. Lati.*

'Which way did she go?' I asked Auntie.

'The usual way,' she said. 'She told me she was going to the office.'

I ran like a madman to my car.

'Is anything wrong?' shouted Auntie from behind me.

But I was gone. I drove furiously to the *Chronicle* offices, and ran straight up the stairs to see the editor, Debo Ojulari. He rose to greet me with a warm but guarded smile, as I quickly scanned my old desk.

'You are most welcome, Mr Dukumo,' Debo greeted. 'Please have a seat.'

'I can't, Debo,' I said. 'I'm sorry. Is Miss Ogedengbe around?'

He cleared his throat.

'She came in here earlier in the day,' he said. 'About nine or a little later, said she must go ... please have a seat and let's talk about the –'

'Go on. Said she must go where? Kwarafa?'

'Yes. Talked about a possible scoop, and demanded a cameraman and a van with a driver. I thought she –'

'And you gave them to her?'

'Well, Mr Dukumo,' he said, nervously now, 'you know Miss Ogedengbe. You know what she's like when she's bent on something. She won't give up until she's got want she wants out of you. And you know she's a damn good reporter. So I had no choice but –'

'You fool!' I spat at him, grabbing his collar. 'Didn't she ...'

I managed to calm myself down and let go of him.

'I'm sorry, Debo. But didn't she tell you she was preg... Never mind. When did she leave?'

'They took off around ten, maybe earlier. What's the matter, Mr Dukumo?'

But I was already on my way, heading back to the Security offices. I can't remember when I got there – I'd driven so fast. When I got there Yelwa was still in his meeting. I gave Suleiman a note asking to see him urgently.

'Sir,' I said, as he came through the door, 'we must act with speed. You remember that right from the start the CCC, to which Harrison and his friends belonged, was constantly harping on about two things – the oil pollution in the Delta and the reduction of the volume of water going downstream through the Kwarafa Dam. Now he has blown up oil installations, I fear he may be planning to blow up the dam.'

'Blow up the dam?' he said, with an expression on his face that was something between amusement and alarm, as though he thought only a madman would embark on such a venture.

'Yes,' I said emphatically. 'I can't explain the bridge. I suspect it's either an act of sheer devilry or a decoy to divert you from his real path. But I fear strongly he might attempt the dam. He can do it, believe me.'

I think he must have come to concede the possibility, for the earlier expression on his face gave way slowly to a worried look.

'We have no time to waste,' I egged him on. 'He may already be there now.'

He looked at me and nodded with growing conviction.

'If you will alert the police command at Kwarafa and then send a platoon from here I'd like to go with them.'

'This is a job for armed men, Mr Dukumo,' he said. 'I don't see how you can operate there.'

'Suppose he gave conditions which the government must meet if it wished to avert the disaster?' I asked with increasing urgency. 'I might be able to talk him out of his plan, but he won't listen to any of your men.'

It was finally agreed. In another hour or so a detachment of armed police boarded an armoured vehicle with their offensive and defensive weapons, and I got in the front with the driver and an officer. They didn't look particularly happy to have me in their company, but they had their orders, and in any case I was too anxious thinking about Lati to bother about their feelings towards me.

The journey from Lagos to Kwarafa is a long one, and so anxious was I that, although we travelled with the speed for which these vehicles are notorious, we were still not going fast enough for me. The driver refilled at Jebba, and it was there that we saw the first sign of trouble. The petrol attendant operated his pumping machine manually, and when he was asked why, he answered that the lights had gone out about half an hour before! I honestly can't tell you, Tonwe, what happened to me from this point on in our drive towards Kwarafa. All I remember is that we had scarcely gone two miles after we turned into the last stretch of road to the place, when we ran into a chaotic rush of vehicles heading in the direction we were coming from. They were preceded by a fire engine with a maddeningly shrill alarm and an officer leaning out of one of the doors and ordering everyone coming from our end to turn back and flee for their lives. The truck slowed down briefly when it got to us, and when our officer asked the fire officer what the matter was, the latter answered at the top of his voice:

'The river! The river is approaching!'

'What?' shouted our officer in disbelief.

'The dam has been blown up! Everything has been swept away by the river, and it is now swallowing up this road with incredible speed. Hurry, turn round!'

Our officer motioned to our driver to turn quickly. With a sudden dash of impudent frenzy I queried the decision, asking the officer if it wasn't better for us to see for ourselves than to depend on the hasty advice of a third party – after all, we had been sent on a mission. I don't remember what the officer said to me, because I was immediately greeted by a wild chorus of angry voices from the platoon of policemen in the compartment behind us, who never welcomed my presence anyway and who, if I'd been foolish enough to tell them my real problem, would have very gladly tossed me out of the van to go and attend to it. Outnumbered and outpowered, I surrendered myself to the inevitable. I don't know if they saw me crying, but I feel certain my tears cannot have mattered to them in the frenzied and irrevocable retreat from our aborted mission.

I must finish up, Tonwe, because the electricity will be switched off in about an hour. Since the destruction of the nation's principal source of power, we are being fed by one of the subsidiary grids in this region. So there is severe rationing of electrical power, though that won't mean much to folks like you in villages where such a facility has never existed. Anyhow, we still watch the news on TV – although the transmission is rather faint, and at the end of the news, about 10 p.m., the lights go off for the rest of the night until 6 a.m. the next day.

At night I keep vigil with Auntie Kumolu-Davies – Judge Benson and a few relatives have been round a number of times – listening to the news and hoping for some sign of Lati. Her parents have not been notified yet, for we are still hoping that she will turn up, in whatever state of health. The NTV crew have been able to bring us horrid pictures of the wreck of Kwarafa Dam and the submersion of villages for miles around, taken from a military reconnaissance helicopter. Five nights ago they showed an interview with two researchers from the National Geological Surveys, who claimed to have been conducting a field test on one of the peaks in the Jebba plateau range when they observed the disaster. According to them, there were about six or seven explosions accompanied by gigantic bursts of flame, which they saw quite clearly from their position. They said the din was frightening and shook the plateau like an earthquake. I can't tell you how much despair that report brought to Auntie and me as we listened. I myself was in no frame of body or mind to console her as she moaned and wailed.

I'm tired, Tonwe, and it's almost time for the news. I've noted what you said about the book – no chapter titles. All right. But the book is not my real problem now. Maybe later. I hope you'll understand.

Piriye

Lagos, 28.2.78

Tonwe,

They've got Bickerbug. I thought you'd like to hear that. No news about Lati, and that makes me even more depressed. But at least I'm glad they've got that godforsaken criminal.

I was with Yelwa yesterday, about 11 a.m. or 12 noon, eavesdropping on reports and phone calls coming to him. I'd finally made up my mind to tell him my girl friend had gone off to Kwarafa without telling me, in the desperate chance that she might catch part of Bickerbug's mad mission. You must understand it took a lot of agonising for me to come to that decision.

Anyway, I was just working myself up to take courage and trying to find the right words to use, when a call came through to Yelwa. I should have told you that since the Lugard Bridge blast, the NSS have put out information on him – complete with pictures – to various units across the country. Well, Bickerbug was smart enough to slip through the various nets thrown around him, but he ran out of luck after his Kwarafa Dam caper. The report that came to Yelwa that morning was from the Idi-Iroko border security – our friend had been trying to slip out of the country, but they got him. They were now bringing him to Lagos under armed escort!

I was tense with anxiety throughout the two hours it took the police from Idi-Iroko to bring in their prisoner. I'd excused myself from Yelwa to go out and get some lunch, but I never did any such thing. Instead I prowled around the premises and the street outside the headquarters, occasionally coming back into the building and hanging around the reception foyer, confronted every now and then by the disdainful looks of policemen who still considered me very much an outsider imposed upon them. Within an hour from when Yelwa got the message from Idi-Iroko, there was a crowd of newsmen on the premises – radio, television and newspaper correspondents, with notebooks and cameras and other tools of the trade. I exchanged courtesies with some of the more familiar ones. I'm not quite sure how they got the tip – they certainly couldn't have been phoned from Idi-Iroko, considering how unreliable the telephone system is in the country generally. Nor would a reporter who observed the event at Idi-Iroko want to let a colleague back in Lagos finish the scoop for him – he'd do everything he could to get to Lagos with all possible speed. You know what I'm talking about.

Anyway, as I said, there was a sizeable crowd of newsmen at the Security Headquarters by this time, waiting to record the arrival of Bickerbug under armed police escort. The Security bosses had obviously been in constant contact with the team that was bringing Bickerbug in. They also knew that a whole party of newsmen were outside there waiting to record the event and create a scene. The news crew had been there for nearly an hour when a navy blue 504 Station Wagon sped into the premises heralded by its siren. There was a stampede as the newsmen struggled to get close to the slowing car. I kept a cautious distance from the scene, standing near the entrance to the building. I was a nervous wreck, but I tried as hard as I could not to show it. Almost as soon as the car came to a stop near the steps leading up to the reception foyer, Yelwa appeared at the doorway in the company of the Inspector-General of Police and two other officers.

The stampede towards the car increased as the doors were thrown open and a handcuffed Bickerbug was led out of the back seat of the car by two policemen. Cameras flashed, but the police held back some reporters who were holding microphones to Bickerbug and firing numerous questions at him. You wouldn't believe what Bickerbug looked like. He had completely shaved off his heavy beard. He had on a white ankle-length cassock, a respectable pair of laced black shoes and a smoke-tinted pair of Polaroid shades over his eyes. But the cassock was stained with dirt, and one of its long sleeves was half-torn from the shoulders. One of the policemen escorting him was carrying a black vinyl briefcase.

I gathered later that Bickerbug had somehow – trust him – acquired a passport identifying him as a Roman Catholic priest from the Cross River State, under the name of Reverend Father Pascal Obongha, on his way to an international theological conference in Coutonou. He had almost evaded detection at the customs check in Idi-Iroko, when one smart official noticed the scar on his forehead and consulted the papers in the security booth bearing Bickerbug's picture with the relevant information on him. When Bickerbug saw the customsman conferring with the security policemen, he turned round and fled from the check-line, briefcase in hand. The police raised the alarm and gave chase. He had actually got as far as a taxi in the garage outside the border post and was holding a knife to the throat of the driver. The driver had started the taxi, but before he could move, the policemen had got to the car, one pointing his pistol to the head of the driver and a second pointing his own to Bickerbug's head. There was a slight scuffle when Bickerbug was ordered out of the taxi, and it was in the process of this that the cassock was messed up a bit.

I hated Bickerbug profoundly as he was being led up the steps of the Security building, amid the stampede of reporters and the restraining policemen. He had a smile on his face as he looked confidently ahead of him, ignoring the numerous questions thrown at him. The crowd had nearly got to Yelwa and the I–G when Bickerbug caught the spiteful look on my face where I stood. He stopped and had the nerve to address me.

'Well, well, well, Piriye,' he said, smiling even more broadly at me, 'we have won, haven't we?'

The crowd parted as he took the policemen leading him a couple of steps towards me. All eyes were on me.

'What nonsense are you talking about?' I returned.

'Our people have won,' he said. 'The water is flowing again, full stream. The tides are here again. Soon there'll be plenty of fishes swimming again, eh?'

The hatred welled up within me and rose to my throat. I couldn't stand the sight of him any longer. As the policemen were steering him towards their bosses, I charged at him in a fury and grabbed his cassock by the collarless neck.

'You bastard!' I screamed. 'Where is my wife? What have you done to my woman?'

Other policemen rushed in and tried to tear me off him, still screaming at the criminal, and cameras flashed all around us.

'What the hell is he talking about?' I heard Bickerbug shouting as the policemen pulled at my hands. 'Would someone get this maniac off me!'

They finally loosened my hands from his cassock, but not before I'd ripped off several buttons down to his stomach, exposing his vest. The policemen took him away. Just before he finally disappeared he turned and called to me:

'Hey, Piriye!'

I looked at him with unmitigated spite, but I didn't answer.

'Don't forget to join me soon, okay?' – and he let out the most exasperating cackle before he was finally led out of sight.

Hours later I regretted that I'd made such a fool of myself there. I didn't stay to observe the rest of the scene. But when they showed the pictures during the NTV news last night I could see that the I–G had declined to make any definite statements in response to the numerous questions posed to him by the newsmen.

Honestly, I don't know what's going to happen next. I'd be lying if I told you I wasn't afraid. In fact, I'm so worried that I haven't eaten anything since I came home yesterday afternoon. The scene I made at the Security building must have got Yelwa and the others curious. What if they've been interrogating Bickerbug in detention, if not before, and he's been telling them far more about my involvement with him than I've confided to them? How safe am I? But I'm much more worried about Lati. I don't want to admit it, but I fear the worst has happened. She is still not back, and the reports in the media continue to increase our despair.

I don't know what else to say, Tonwe. Besides, the news comes on in about twenty minutes, and then it's lights out. I'm frightened and I'm depressed. Every time there's a knock on Auntie's door, I think they're looking for me. But I really have no wish to run or hide. Who knows – maybe I'm wrong!

Priboye leaves tomorrow. I trust you'll find some way of making contact with me, if he's going to be occupied in Warri for as long as he says. *Please* write to me. You can't know how much reassurance your letter will bring to me.

Piriye

P.S. It's just been reported on the NTV news that some survivors from the Kwarafa disaster are receiving attention at the Jebba General Hospital. No names were given, but we can't help being hopeful. I'll be travelling to Jebba first thing tomorrow morning. Oh Tonwe, *please please* pray for us!

Piriye

Piriye –

This is Priboye. I cannot say much. Cannot even tell you where Im writing from. Have been hiding from one house to another in this town. The police are after me. You may be in trouble yourself. Please run. Remove yourself from Lagos. Until the situation becomes clearer. I went Seiama 2nd March deliver your last letter. Did not find Brisibe at home. His wife was there. When she saw me started crying. Some women there too. They tried console her. I asked what's matter. She said husband was in police custody. One of those arrested over explosions had talked police. Said he had visited Brisibe in company of Harrison. Police went confront Brisibe. What he have to do with Harrison gang? He said had nothing to do with them. Said had sent them out his house. Why did not report visit? Brisibe answer didnt satisfy them. So they arrested him. Came back later search his house. Took away lot documents. Mrs. Brisibe said your letters among. Husband arrested about end last month. She hasnt seen him since. They wont tell her where he is. I returned Warri with your letter. Its here with me now. 6th March went to my construction site at Eneren. My contractor told me two policemen come to ask for me. What did I have do with the police? Something told me their visit to do with Brisibe. I did not stay long there. Later in evening my wife told me two strange men come looking for me. About 5 p.m. I didn sleep my house that night. Packed small case. Spent night with friend on Cemetery Road. About 6.30 next day my friend ran into room. Said two men asking for me in front yard. I quickly packed my things. Ran out from back house. Did you mention my name in your letters to Brisibe? That was big mistake. Havent seen my house or office four days. Im getting tired of running. What have I done but carry letters for my friend? Please run. Im sure trouble is connected with contents of letters. My host says his brother leaves for Lagos early tomorrow. Its 9.15 p.m. now. Must give this letter and yours to him. Please take my advice. Find somewhere hide. Until this thing settles. Must stop at once. I can hear somebo

Other titles in the Longman African Writers series:

Heroes Festus Iyayi
Flamingo Bode Sowande
Sugarcane with Salt James Ng'ombe
Hurricane of Dust Amu Djoleto
Man pass Man *and other stories* Ndeley Mokoso
Loyalties *and other stories* Adewale Maja-Pearce

Titles in the Longman African Classics series:

The Park James Matthews
The Cockroach Dance Meja Mwangi
The Beggars' Strike Aminata Sow Fall
Flowers and Shadows Ben Okri
Violence Festus Iyayi
Call Me Not a Man *and other stories* Mtutuzeli Matshoba
The Children of Soweto Mbulelo Mzamane
Hungry Flames *and other stories* Mbulelo Mzamane
A Son of the Soil Wilson Katiyo
The Life of Olaudah Equiano Paul Edwards
The Stillborn Zaynab Alkali
Muriel at Metropolitan Miriam Tlali
The Marriage of Anansewa *and* Edufa Efua Sutherland
Native Life in South Africa Sol T. Plaatje
The Victims Isidore Okpewho
The Last Duty Isidore Okpewho
Fool *and other stories* Njabulo S. Ndebele
Sundiata: an epic of old Mali D T Niane
Master and Servant David Mulwa
The Dilemma of a Ghost *and* Anowa Ama Ata Aidoo
No Sweetness Here Ama Ata Aidoo
Our Sister Killjoy Ama Ata Aidoo
Scarlet Song Mariama Ba
Tales of Amadou Koumba Birago Diop